W9-ATY-701
3 1299 00966 0466

East Meadow Public Library
1886 Front Street, East Meadow, NY 11554
(516) 794-2570
www.eastmeadow.info

TED Books

# Why Dinosaurs Matter

KENNETH LACOVARA

ILLUSTRATIONS BY MIKE LEMANSKI

TED Books
Simon & Schuster

New York London Toronto Sydney New Delhi

Simon & Schuster, Inc.
1230 Avenue of the Americas
New York, NY 10020



First TED Books hardcover edition September 2017



For information about special discounts for bulk purchases, please contact Simon & Schuster Special Sales at 1-866-506-1949 or business@simonandschuster.com.

For information on licensing the TED Talk that accompanies this book, or other content partnerships with TED, please contact TEDBooks@TED.com.

Interior design by: MGMT. design
Jacket design by: MGMT. design
Artwork by: Mike Lemanski

Manufactured in the United States of America

10 9 8 7 6 5 4 3 2 1

Library of Congress Cataloging-in-Publication Data is available.

ISBN 978-1-5011-2010-7

ISBN 978-1-5011-2011-4 (ebook)

*For Jean*

*I'm a lucky man to share this skinny slice of deep time with you.*

CONTENTS

# Why Dinosaurs Matter

# 1 In Defense of Dinosaurs

Albert Einstein was a complete and utter failure. His name should be synonymized with obsolescence. The mere mention of his person should connote a devastating inability to adapt to change. Of course, it's true that he revolutionized science, invented our current framework for understanding the cosmos, and bent our very perception of space and time. He won the Nobel Prize in 1921, was named *Time*'s Person of the Century, and was awarded honorary doctoral degrees from Oxford, Princeton, and Harvard Universities. The principles he laid down made possible the development of GPS, digital cameras, smoke detectors, burglar alarms, cell phones, and countless other consumer products. Computers and semiconductors themselves would not be possible were it not for Einstein's March 1905 paper setting forth his particle theory of light, cracking the bedrock of established doctrine and laying the foundation of modern physics. Just a few months later, Einstein cinched the case for the existence of atoms, silencing the gainsayers on the other side of an age-old debate. Arguably, the modern age, as we know it, would not have come to be or would have been delayed without the accomplishments of Einstein.

Yet where is Einstein now? Dead, that's where. Despite his stunning intellectual prowess, his ability to completely disrupt and replace our view of the cosmos, and his heroic achievements that improved the lives of every human to follow, Einstein, in the end, died. A catastrophic event befell this most brilliant and successful man—an abdominal aortic aneurysm. Unable to adapt to the sudden and dramatically changing physiological conditions within his own body, Einstein departed life at age seventy-six, doomed by his inability to overcome the devastating angiological circumstances that led to his demise. Having perished, having succumbed to his own mortality, having vanished from the living Earth, we must now conclude, through the lens of history, that Einstein's legacy is a cautionary tale about failure to adapt and about eventual obsolescence. He died and was, in his very last moment, unequivocally, a failure.

Preposterous! Ludicrous! Of course. It would be the height of absurdity to conclude that Albert Einstein's towering accomplishments should in any way be sullied by his own mortality. Despite a lifetime spent blazing trails across the frontiers of human understanding, his last step was the last step we all must take: one final footfall over the threshold of life, into oblivion. Einstein was a great man, but a man nonetheless—a human, a *Homo sapiens*—every one of which lives for some short while and then dies. It's what we organisms do. Marie Curie, Benjamin Franklin, and Charles Darwin are no less great because they eventually succumbed to disorder and died. Louie Armstrong will forever remain the founding savant of jazz, even though his horn has fallen silent. Neil Armstrong's "giant leap for mankind" will never be walked back. His indelible tracks will endure in the dust of the Moon and in the minds of humans for as long as those vessels persist to bear the trace of his accomplishments. To argue otherwise, as I have done above, is ridiculous in the extreme.

Now that we've dispatched the preceding deeply flawed line of reasoning, I'd like to pose a question: Why do we besmirch the legacy of the dinosaurs using the very same foolish argument? Why is the word *dinosaur* used so often as a pejorative to describe obsolescence? How did *dinosaur* become an epithet to invoke an inability to adapt to changing conditions? Why are dinosaurs the one group of animals associated most closely with failure?

These slanders against the good name of the dinosaurs are no mere colloquialisms. They are proper English. They have been codified, legitimized, and set to type by the lexicographers of all the major English dictionaries. Crack open the Merriam-Webster dictionary, and you will find that the word *dinosaur* means "one that is impractically large, out-of-date, or obsolete."[1] Consult the Cambridge Dictionary and learn that a dinosaur is "something that is old and that has not been able to change when conditions have changed and is therefore no longer useful."[2] Heft open the queen mother of all English lexicons, the Oxford English Dictionary, and you will find definitive proof that *dinosaur* can be quite properly used to refer to "someone or something that has not adapted to changing circumstances."[3]

With such weighty authority against them, it's no wonder that the legacy of the dinosaurs is so often dragged through the metaphorical mud. I could fill a book with defamatory comparisons to dinosaurs. Here's a brief sampling: IBM is an "IT Giant Commonly Viewed as a Dinosaur," the headline read. "Intel: A Dinosaur Headed for Extinction?" pondered an investment site. "Both Major Parties Are Seen as Dinosaurs—Old Institutions That Do Not Fit the Times or Challenges of the Day,"[4] opined the *Wall Street Journal*. To all that, I say humbug! They should all hope to be so lucky.

What CEO wouldn't daydream lustily about global dominance spanning a geological era? What board chair wouldn't crave the rapid growth of thousands of successful franchises the way that dinosaur species exploded across the globe, as they conquered continent after continent? What head of R & D wouldn't revel in the development of unprecedented feats of speed and size and power and versatility? Dinosaurs pushed the envelope of physiological possibility, broke record after record, and were paragons of success by almost any measure.

Considering, as a whole, the breathtaking adaptations of dinosaurs, such as titanic size, devastating power, extravagant plumage, razor-sharp teeth, and bizarre spines, plates, horns, and clubs, the public adoration for these amazing creatures is not surprising. What is surprising is our dichotomous relationship with the concept of dinosaurs. How did these versatile creatures, arguably the most successful group of large land animals in Earth history, get labeled as the epitome of prehistoric failure?

The most damning misconception about dinosaurs is the idea that their extinction, except for the birds (more on that later), represents their own failure to adapt to changing conditions. Until relatively recently, the idea seemed patently true. If only they weren't such dim-witted, sluggish, stuck-in-the-mud, ponderous creatures, maybe they could have survived and hung on to the domain that was once theirs. But they weren't good enough. Not clever enough. Not adaptable, like our own tiny ancestors. And in the end, they couldn't hack it, and the cream, like it always does, rose to the top. The mammals took over, and here we are, smarty-pants primates, with dominion over the Earth. That was the narrative.

Inculcated for decades with a host of misconceptions, it's easy to see how the public came to view these noble and spectacularly successful creatures as failures: as the evolutionary equivalent of the VHS tape, the DeLoreans of the Mesozoic Era,[5] the Woolworths of the fossil record. Prior to 1980, no one had any idea what had happened to the dinosaurs. They were here for 165 million years, and then—*poof!*—they were gone, thought to have petered out in a whimper, perhaps, or maybe something more dramatic. Crackpot theories abounded: A dinosaur pandemic killed them. Those sneaky mammals ate all their eggs. If sex determination in dinosaur embryos was temperature dependent (we have no reason to think this), maybe the climate became too cold to have males or too hot to have females. Their shells got too thin. Caterpillars ate all their food. Maybe the radiation from a supernova got them. Maybe all the fiber-rich plants died out, and they all perished of constipation. Maybe they were just too dumb to live.

If you were born before about 1990, this was the dinosaur narrative of your youth. Dinosaurs were losers, a plotline of extinction by ineptitude drilled into our subconscious by children's books, television shows, and movies. Walt Disney, in his 1940 animated classic *Fantasia*, presented the most gorgeously illustrated version of this sad tale. His retelling of natural history was set to Igor Stravinsky's *Rite of Spring*, which has wildness in it, with its crashing rhythms breaking like surf, and unexpected shrieks and wails—the perfect soundtrack to the Darwinian struggle to survive. Beginning with the violent and fiery origin of the Earth, Disney's version takes us through the emergence of single-celled life, adorable and personable in his hands, and on to the demise of the dinosaurs. The original concept was to continue the story through to human evolution, but Disney scuttled this plan to avoid provoking the ire of creationists.[6]

Disney's depiction of Earth history was guided by some of the top minds in 1930s science, including leading paleontologists Roy Chapman Andrews and Barnum Brown, the biologist Julian Huxley, and the astronomer Edwin Hubble. The result was state-of-the-art. Even now, nearly eighty years later, portions of *Fantasia*, such as the emergence from the sea of limbed animals and the evolution of flight, seem quite modern. And, of course, the climactic *T. rex–Stegosaurus* death match is a cinematic classic. (Never mind that *T. rex* and *Stegosaurus* were separated by more time than separates *you* and *T. rex*.) "Don't make them cute animal personalities," Disney cautioned his animators. "They've got small brains, y'know; make them real."[7] And real they were. The most realistic dinosaur reconstructions the world had seen. The *Rite of Spring* scene in *Fantasia* thrilled audiences. It was the *Jurassic Park* of its day.

Disney wraps up his natural history with a somber epilogue; a requiem to the demise of the dinosaurs. Climate change, it seems, is at work. We're led to believe that volcanoes belching gas into the sky have despoiled the Cretaceous[8] atmosphere. Plants wither under the blazing orange sun, water evaporates, and the ground becomes parched and dusty. Thirsty dinosaurs in search of water plod stiffly, like zombies, through a hellacious landscape. They paw and nose feebly at a dried-out watering hole, but no water awaits. One by one, they succumb to dehydration and flop morbidly across the cracks of a desiccated lake floor. Unable to adapt to the changing world around them, hampered, perhaps, by their small brains and indolent nature, the dinosaurs perish. Disney leaves the final four saurian protagonists trudging to their doom as a dust cloud envelops them. Pan to the searing sun, set into a baked orange sky, and fade to black. The Cretaceous is over. Although Disney chickened out on the last chapter, and the scene ends there, the implication is that with the unadaptable dinosaurs dead, a new world awaits our worthy heroes—the mammals.

In this, and in a thousand similar accounts, the dinosaurs of our imagination were tarred with fatal incompetence. There was nothing to learn from prehistory's biggest losers, except: Don't be like the dinosaurs. Don't be a failure. Don't be incompetent. Don't be obsolete. And with that, the long history of success enjoyed by the dinosaurs was turned on its head. The descendants of the tiny fuzzballs that huddled in fear in the dark and little-noticed recesses of the dinosaur world had grown up and labeled their former tormentors stupid and irrelevant. The label stuck. But the bum rap, hung about the necks of the dinosaurs, reflects our failure, not theirs. It stems from our long-running inability to decipher the true nature of their extinction.

Geological literacy is new to humanity. Like a child learning to read, we cracked open the rock record and saw to our untrained eyes a confusing jumble of letters and words—alphabet soup carved in stone. By the time dinosaurs were recognized as a group apart from other animals, we could sound out a few words and grasp the major themes of Earth history, but the nuances of our planet's story were beyond our apprehension. Dinosaurs were here, and then they were gone; we could see that. But the nature of their disappearance remained a total mystery until 1980, when Luis and Walter Alvarez, father and son, presented evidence[9] that the dinosaurs did not fail to thrive, vanishing from the world of their own degeneracy. They were murdered. Snuffed out by a space rock that unleashed hell on earth. It took decades for this idea to catch on, particularly with paleontologists, but most now seem to have finally come around to it.

Dinosaurs, exculpated from blame in their own extinction, should no longer bear the tarnish of failure. Their demise should no more diminish their achievements than Einstein's death should abate his. They were, and still are, an unqualified success. Their name is clear. They died through no fault of their own. What's more, we can learn from them. We should learn from them. To do otherwise would be foolish and arrogant.

Dinosaurs are long-lasting champions of resilience and persistence. They reigned unchallenged on the land for the better part of 165 million years. But that's only if you exclude birds, which are truly dinosaurs. If you include the birds, known now as avian dinosaurs,[10] their incredible run has yet to pass and spans the last 231 million years. Not bad. Primates have been around for about 56 million years.[11] Our human lineage split from the line leading to chimpanzees only six to seven million years ago.[12] Our own species appeared around 200 thousand years ago.[13] That's a rounding error on the Mesozoic timescale.

Perhaps it is fairer to compare all mammals and all dinosaurs (avian and non-avian). Our forebearers, tiny shrew-like creatures with the unlyrical name morganucodontids, first appeared about 210 million years ago.[14] That's a respectable run across deep time. But by the time of their first appearance, dinosaurs had already walked the Earth for twenty-one million years. If every bird on every continent were to die today, mammals would not surpass the temporal success of the dinosaurs until the year 21,002,017.[15]

Dinosaurs are ancient *and* contemporary. The latest chapter of their triumphant reign is being written as you read this. Across the blinding ice sheets of Antarctica, through the cacophonous forests of Amazonia, atop the withering heights of the Himalayas, in oases dotting the Sahara, and in a million other places—including your backyard bird feeder—the exquisite adaptations and enduring persistence of the dinosaurs is on display.

As biologists work to expand their understanding of modern avian dinosaurs, growing ranks of paleontologists are adventuring to the far-flung corners of the globe and uncovering ancient dinosaurs at an ever-increasing pace. From their discovery, in the early nineteenth century, until the mid-twentieth century, the recognition of new dinosaur species was a rare event, only about one per year. By 1990, the rate had climbed to about six per year. By 2006, it was fifteen. In 2016, thirty-one new dinosaur species were described! Every year, we discover that dinosaurs were more widespread, more diverse, and more amazing than we ever dared to imagine.

This treasure trove of information comes to us at an auspicious moment. As we move into an uncertain environmental future, it has never been more important to understand the past. The lessons written in stone and buried beneath our feet are both profound and urgent. Dinosaurs are more than the charismatic stars of deep time. They are the victors of a ruthless selection process that dealt death to any species that faltered. They are the products of a trillion experiments played out according to the uncharitable rules of evolution, and we have so much to learn from them.

Want to design a system to move heavy loads over rough terrain? Dinosaurs did that. Want to understand mostly passive and efficient cooling systems? Sauropods were experts. If you're interested in extremes, dinosaurs include the largest creatures ever to walk upon the planet and, also, nearly the smallest. Interested in upcycling, in repurposing technology? Look to the dinosaurs. Feathers, in particular, are a marvelous example of exaptation: the process of acquiring functions for which they were not originally adapted. Looking for the functionally important segments of a protein? A 2011 study, guided by earlier work in molecular paleontology, found that the most crucial regions of the collagen molecule, the most abundant protein in your body, are those that are preferentially preserved in the femur (thighbone) of a *Tyrannosaurus rex*.[16]

Want to understand market segmentation and capture? The dinosaurs, from tiny hummingbirds to sixty-five-ton plant eaters, are champions of niche partitioning. Want to explore adaptation? Today avian dinosaurs are the only large

Until now, you might have seen dinosaurs as cool comic-book characters, all gnashing teeth and thundering limbs, celebrated in children's fantasies and blockbuster movies but otherwise quaintly obsolete. The truth is far different.

Dinosaurs were finely tuned marvels of evolutionary adaptation. Old as they are, they are supremely relevant today. To study them is to be astounded by their physical virtuosity—and to gain valuable insights about endeavors as varied as flight, industrial design, and locomotion.

creatures to have breeding populations on every continent. Interested in resilience? Avian dinosaurs survived the worst catastrophe in last quarter billion years and today outnumber mammalian species by more than three to one. Since da Vinci, and probably long before, humans have been fascinated with self-powered flight—something that we've been unable to substantially achieve. Dinosaurs did this 150 million years ago.

Hardly the embodiment of obsolescence, more than eighteen thousand species of avian dinosaurs[17] flit, trod, and swim about our planet today. To be a dinosaur is to belong to a staggeringly successful group of animals whose reign across time may never be matched by humans or any of our mammalian kin.

Beyond their admirable qualities, dinosaurs resonate with the public in special ways. They embody the past and stand in for the ancient. But why look back, when so many challenges lie ahead? In this book, I will argue that dinosaurs matter because our future matters. Global warming, sea level rise, the catastrophic degradation of our environment, and the heartbreaking and costly biodiversity crisis all loom large on our horizon. People, even paleontologists, are more concerned with the future than with the past. But we don't have access to the future. We can make no observations of it and can conduct no experiments in it. The future is a dark scrim that races just before us, always obscuring that which we are about to experience, always concealing how the world will dispose of our dreams and hopes and prayers and desires. As for the present, there's not much to it. Unstable and fleeting, like the heaviest of elements. A wisp of time separating that which can be from that which has been. The sentence you are reading is already in your past. But the past can be embraced. It's in the hills, under the oceans. You can hold it. Crack it open. Learn from it. Put it in a museum for all to see. Most importantly, the past is our guide to the future, the only one we will ever have.

When the infamous bank robber Willie Sutton was asked, "Why do you rob banks?" he replied, the story goes, "Because that's where the money is." Why study the past? Because that's where the answers are. If you're concerned about the multiple, simultaneous existential crises facing humanity, look to the past. No analogies are perfect, and the ancient record does not contain all the answers. But we would ignore it at our peril. Winston Churchill is quoted as having said, "The further back you look, the further ahead you will see." Only the past provides the broad view that we desperately need to prepare for the future. We can now gaze out upon Earth's ancient worlds from many vantage points; vistas cut through time for the passage of our imagination. Each view has much to teach us, but no lookout commands the attention and holds the allure of the precipice on which we stand when we view the ancient world of the dinosaurs.

# 2 Is a Penguin a Dinosaur?

The sunlight filtering in through the aluminum venetian blinds lent a film noir quality to the rooms of our little home on the bay, across from Atlantic City, New Jersey. I sat on our slipcovered couch, transfixed by the grainy black-and-white image on our small RCA television. Two men were inside a rocket—a real rocket—and they were being launched into space. I was probably four years old. By my best calculation, I might have been watching the March 23, 1965, launch of Gemini III, which hurled astronauts Gus Grissom and John Young into three low orbits around our planet. This scene forms my earliest memory, and every moment of the rest of my life has been touched by the excitement I felt that day for science and exploration.

My older brother, Tom, had begun to bring back fern fossils and quartz crystals from trips to our uncle's dairy farm in Pennsylvania. They were mysterious and fascinating objects to me, and I spent many hours examining them, mostly when he was off playing baseball or fishing in the salt marsh. Then, when I was seven years old, a local rock hound named Mrs. Osler brought shoe boxes full of rocks and fossils into our Cub Scout meeting. I was astounded by what I saw.

The shapes, the colors, the sparkling facets, the animals trapped in stone—how could this be? I was hooked. The next day, no doubt referencing my Golden Guide to *Rocks and Minerals*, I wrote an essay about igneous, metamorphic, and sedimentary rocks. In it I declared my intention to become a geologist, and I opined that sedimentary rocks are the best kind of rocks because you can find fossils in them. (I was right about this, it turns out—they *are* the best kind of rock.)

My route to college and, eventually, paleontology, was not a straight one. After a pretty serious detour into music, during which I toured the country playing drums and ended up pulling a stint as the house drummer at the Golden Nugget casino in Atlantic City, I landed in graduate school, earning my doctoral degree in geology from the University of Delaware.

I began my academic career as a professor at Drexel University, in Philadelphia, where I taught courses in geology, paleontology, and evolution. In 1999 I joined a group of colleagues from the University of Pennsylvania to prospect for dinosaurs in the remote and desolate Bahariya Oasis, in the Egyptian Sahara. We spent two winters there searching for the so-called lost dinosaurs of Egypt: four species of dinosaurs that had been discovered in the desert nearly a century before, taken to Germany, and then destroyed in an Allied bombing raid during World War II. The lost dinosaurs eluded us, but in the course of looking, we discovered a new type of giant plant-eating dinosaur that we named *Paralititan*, meaning tidal giant—a reference to the ancient mangroves that served as its vast salad bowl.

My writer friend Jeff Blumenfeld is fond of saying, "An expedition is an adventure with a purpose." Over the next decade, I found myself around the world on many adventures, each with the purpose of revealing another page or two of the saga that is Earth history. In Egypt, a Bedouin tribesman who mistook me for a grave robber threatened me with a scimitar. (An honest mistake, really.) I came perilously close to falling to my death, in Montana, while clambering down a cliff face, with a rock strapped to my back that preserved the tracks of Jurassic[18] pterosaurs. In China, while on the rickety night train to Jiuquan, I nearly puked out my internal organs after contracting the worst case of food poisoning I've ever had. (In my house, "taking the night train to Jiuquan" is now the standard euphemism for being sick.) In the Gobi Desert, I was attacked by a homicidal Bactrian camel and had to run for my life. I've had a pit viper curl up between my feet, scorpions living under my tent, pumas snooping around my camp, rattlesnakes rattling at me, and bulls chasing me, more than once. (My most recent encounter with a grumpy bull occurred on the North Sea coast of Scotland, while researching this book.) In time, I would find myself rafting down a glacial stream in southern Patagonia, Argentina, where I would eventually happen upon the grave of one of the most magnificent titans to ever walk this planet. Nine years later, I would name this massive plant-eater *Dreadnoughtus*, the dinosaur that feared nothing.

Although paleontology is my everyday profession, I remain astonished that there were dinosaurs in the world. And I feel lucky to be alive now in an age in which we recognize fossils for what they are, an age in which we can explore the ancient past like never before. Beyond dinosaurs, the fossil record is bursting with mind-blowing creatures. Sea monsters, like mosasaurs, were real. *Quetzalcoatlus*, a pterosaur that was as large as a Cessna and stood as tall as a giraffe, was real. There were filter-feeding crocodiles and whales with legs. There were rodents the size of cows, and snakes that could eat cows. There were dragonflies with the wingspan of an eagle and pill bugs the length of a car. There was a Cambrian[19] animal that looked like a cross between a gummy worm and a box of toothpicks, a creature so bizarre that paleontologists named it *Hallucigenia*. The bounty of wonders is endless, and nature never fails to surprise.

When I give public talks, I sometimes engage the audience in a fun game of Which One of These Is a Dinosaur? I'll show images of four animals and ask the audience to shout out which ones are dinosaurs and which ones are not. The images I show are of a mosasaur (a giant marine predator that lived during the Mesozoic), a pterosaur (a large flying creature, also from the Mesozoic), a crocodile, and a cute, fuzzy penguin. Most people believe the mosasaur to be a dinosaur. It's huge and scary and has *saur* in its name. Almost everyone votes the pterosaur into the ranks of the dinosaurs. Pterosaurs appear in nearly every children's book about dinosaurs, and their name also contains the clue *saur*. Most people think the crocodile is not a dinosaur, but it usually gets a few votes. Then the fun part: "Is a penguin a dinosaur?"

At this point, the room usually erupts into laughter and a cacophony of opinions. If I'm doing this bit at a school, the teachers will usually have to step in after the penguin question to quiet things down.

The answers:

Mosasaur—not a dinosaur. Mosasaurs were marine lizards, more closely related to the Komodo dragon than to dinosaurs. How do we know this? They do not possess the defining anatomical features of dinosaurs.

Pterosaur—not a dinosaur. Pterosaurs, which include the famous pterodactyls, were flying reptiles that lived with the dinosaurs but are not dinosaurs. Again, they do not possess the requisite anatomy. Pterosaurs are closely related to dinosaurs but branched from a common ancestor before the first dinosaurs appeared.

Crocodile—not a dinosaur. Crocodiles are the closest living relatives of dinosaurs but are not dinosaurs. They also fail the anatomy test.

Cute, fuzzy penguin—dinosaur! Penguins, along with all birds, are dinosaurs. They are not closely related to dinosaurs. They *are* dinosaurs. Being a dinosaur is a binary condition; there are no degrees of being a dinosaur. Either you are a dinosaur, or you're not. Avian dinosaurs, as we call birds, either possess, or have ancestors that possessed, all of the defining characteristic of dinosaurs. Birds are dinosaurs in the same way that a *T. rex*, a *Stegosaurus*, and a *Dreadnoughtus* are dinosaurs.

That sounds crazy, you might be thinking. Indeed it does, but that's why we have science. As history demonstrates, common sense is a poor guide to understanding the structure and complexity of the universe. Albert Einstein called it "a collection of prejudices." If common sense worked well, we wouldn't need science. But it doesn't. Calling a penguin a dinosaur certainly violates common sense, but it's true.

So why is a penguin a dinosaur? We can see in their bones unmistakable clues from their Jurassic forebearers, dinosaurian traits passed forward along an unbroken chain that spans a hundred and fifty million years. Owing to the penchant of some bird species to favor watery environments, in which individuals are occasionally preserved in exquisite detail within the silty bottoms of lakes, avians have left behind a wonderful account of their branching from dinosaurian stock.

Even though we can see that birds evolved from dinosaurs, should we still call them dinosaurs? At some point, shouldn't we say they stopped being dinosaurs and became just birds? No. When our ancestors evolved into humans, they didn't stop being primates. They were not stripped of their membership in the club of mammals. Nor was their animalness revoked.

They remained all of those things. A penguin is a bird *and* a dinosaur, by virtue of the fact that it has dinosaurian ancestors. Like all dinosaurs, birds trace back their lineage to the first dinosaur and are, therefore, dinosaurs.

The key to establishing a penguin's place on the tree of life is to determine the branch on which it rests. Branches can have branches, of course, and a tiny twig of a branch, like a penguin species, is nested on a series of larger branches. A complete branch, which includes its point of departure from the rest of the tree and all of its subbranches, is called a clade. In practice, a clade consists of an ancestral organism and 100 percent of that organism's descendants—the whole branch. Clades are nested in clades, just as crows are found in the clade *Corvus*, which is nested in the clade Aves (birds), which is nested in the clade Dinosauria.

If we took only a portion of the ancestral dinosaur's descendants and called them dinosaurs, and called the remaining portion birds, we would no longer have a dinosaur clade—that is, we would no longer have a complete branch. A useful analogy can be made with your personal family relationships. For example, imagine your great-great-grandmother. Let's call her Polly. Her genetic family, going forward, consists of her and all of her descendants. (Species do not have in-laws, so let's ignore the confounding effects of in-laws in this family analogy.) You can think of your great-great-grandmother's family as a tree, growing upward, with her at the base. Your great grandparents, grandparents, and parents all form their own branch, each sprouting from the preceding one. You are a member of each. Now imagine your first cousin Boris. His great-great-grandmother is also Polly. He's in the family, just as you are, by virtue of the fact that he descends from Polly. It turns out, though, that he's an obnoxious, insufferable bore. So let's kick him out. What? We can't do that. Boris is in the Polly family to the same extent that you are, that is to say, 100 percent. His breath might smell. He might have voted for *that* guy. He might be a bloviating, pontificating, color-clashing, children-frightening ne'er-do-well, but he's in the Polly clade, same as you. You can tell him, "Thanksgiving has been canceled this year," unfriend him on Facebook, and change your phone number to 555-something, but there can be no objective basis for kicking him out of the Polly family.

Now let's create a definition for dinosaurs, based on clades. We will call the biggest branch Dinosauria, and it sprang from the tree of life around 231 million years ago. If we take the ancestral dinosaur and 100 percent of its descendants, all of the branches lead back to the same point: back to the very first dinosaur. Within the dinosaur clade, there are many nested clades. On one side, there are horned dinosaurs, and armored dinosaurs, and a group of plant eaters that are mostly duckbills. On the other side sits a clade that includes the supergiants, the sauropods. Next to it sits the clade containing the meat eaters, like *T. rex* and *Velociraptor*. Around 150 million years ago, this group gave rise to a subclade of feathered dinosaurs. They looked much like their non-avian feathered dinosaur cousins, but they could fly. The birds had arrived.

Whether avian or non-avian, all dinosaurs possess a characteristic set of anatomical features that sets them apart from their reptilian ancestors, as well as from their living and extinct cousins—a set of features that is the essence of dinosaurness. They all pertain to enhanced power and vigor.

A suite of novel innovations allowed dinosaurs to engage in a more energetic, more kinetic mode of being, compared to their precursors. In particular, dinosaur limbs are all about strength and forward motion. Unlike their predecessors, dinosaurs held themselves in an upright stance. Whether flesh rippers or plant gobblers, they were dynamic creatures whose bodies were held at the ready, poised for action. Dinosaur limbs were situated under their bodies, nearly perpendicular to the ground—more like a horse than like a crocodile. Their rear feet possessed modifications in the ankle that limited side-to-side motion but created a strong forward-facing hinge mechanism, allowing for rapid and efficient straight-ahead locomotion.

Dinosaurs had particularly powerful legs, attached firmly to rigid hips, that served as solid scaffolding for extra beefy leg muscles. Indeed, the strongest legs to have ever been have all been attached to dinosaurs. Dinosaurs were ready to go, evolved to engage in rapid and sustained forward movement.

Now contrast the dinosaurian stance with the languid posture of lizards or crocodiles: limbs sprawling outward, knees and elbows bent, feet splayed to the side, belly scraping the ground, and tail dragging behind. Not an action pose. Quite the opposite. From an energy standpoint, this outspread stance is very conservative—a good strategy for ectotherms who eat infrequently and burn calories parsimoniously. These listless, cold-blooded beasts are always about half a push-up away from taking a nap. Their indolent nature, though, should not be confused with their ability to move rapidly. Many reptiles are capable of attacking prey with blinding speed and of executing astonishingly athletic maneuvers. These are exceptional moments, though, in the reptile day. Most of their time is spent resting. The economy of cold-blooded physiology dictates a high proportion of torpor to vigor.

Once your mind is tuned to look for the requisite ready-for-action dinosaur anatomy, you begin to see the *Velociraptor* in the turkey. With a little practice, you'll start to notice the inner dinosaur in every bird, from a penguin to a pigeon. It may violate common sense, but we can no more whisk a flitting hummingbird off the dinosaur branch than we can boot abominable cousin Boris from the Polly family. Like the Mafia or the CIA (the stories go), once you're in, you're in. Birds are dinosaurs because their ancestors were dinosaurs.

Like a tree that is forked at its base, the dinosaur clade split soon after its inception into two mighty trunks that would each bear many branches and twigs. One trunk is called Ornithischia, and from it sprouted the duckbill dinosaurs, the horned dinosaurs, and the armored dinosaurs. The other stout trunk, Saurischia, branches into the theropods—often called the "meat eaters"—and a group of long-necked, long-tailed plant eaters called the sauropodomorphs. Most members of the latter group are in a subclade called Sauropoda. These are the true giants, the largest animals to ever walk the Earth, such as *Brontosaurus*, *Argentinosaurus,* and *Dreadnoughtus.*[20]

Across the dinosaur clade, the evolutionary plasticity of body size and shape is truly a wonder of nature. Members of the group span an astounding range of sizes, from the 65-ton *Dreadnoughtus* to the bee hummingbird, which barely tweaks the scale at 0.056 ounces (1.6 grams). Remarkably, both species reside on the saurischian side of the dinosaur tree, meaning that the bee hummingbird is more closely related to *Dreadnoughtus* than *Dreadnoughtus* is to *Stegosaurus, Triceratops*, *Pachycephalosaurus*, or any of the other dinosaurs on the ornithischian trunk. This works with any pair of saurischian dinosaurs: a *T. rex* and a flamingo, *Brontosaurus* and a blue jay, and a *Velociraptor* and a ring-necked pheasant are all more closely related to each other than they are to any dinosaur on the ornithischian side of the tree, such as *Triceratops.*

The surprising relationships that exist among dinosaurs only add to their wonder. And the anatomical themes that carry through the group reveal order and beauty, based on motifs, repetition, and variation.

But without a deep-time perspective, we could never make the connection between a hummingbird and a *Dreadnoughtus.* Our human lives are so brief that the unrelenting forces of evolution are nearly invisible to our unaided senses. But just as we use telescopes and microscopes to view what we could never see with our naked eyes, the rock record provides a lens that reveals the slow-motion workings of the world. Viewed this way, the Earth is alive with creative and destructive forces. Tectonic plates race about its surface, scraping along, pulling apart, and crumpling together to form vast ranges of mountains. The oceans lap on and off continents, and glaciers pulse away at the poles like a twin pair of icy hearts.

Viewed over deep time, the tree of life is a dynamic shape-shifter, mostly growing and expanding but occasionally losing a branch or a twig. Its sapling years are largely mysterious to us: the eon before hard-bodies committed story to stone.[21] The half billion years since have been mostly good. But five times over this span, the tree has been severely pruned. Poisoned, frozen, heated, and set to flame, sometimes in combination, the tree of life has weathered five mass extinctions that served up devastation and opportunity in large measure. When the fifth calamity struck, this time from space, the mighty limb that was the dinosaurs was nearly severed. And it would have been, were it not for the tenacity of a single lineage that managed to hang on: the branch we call the birds. Cantilevered way out near the end of this branch sit seventeen species of living penguins, seventeen penguin dinosaurs, seventeen soggy, blubbery examples that science is our best tool for overcoming the collection of prejudices that each of us brings to the natural world.

# 3 Walking Museum of Natural History

The bizarre relationships revealed in the fossil record certainly are not limited to those among the dinosaurs. The pages of Earth history hold plot twists that might garner literary awards, were they conjured from the imaginations of novelists. But they were not. The Earth's story writes itself.

The latest page of this epic—the one we call the present—has an air of inevitability, but it is the product of innumerable contingencies that lined up in just the right way to create the world that we know. It could have been otherwise. Disturb something here, delay an event there, reorder a single step in a long sequence, or shift a continent this way or that, and Earth history changes forevermore. A single sunbeam causing a single mutation is all that it takes. A space rock nudged an infinitesimal degree to the left or right could change the course of all it is yet to pass. Kill off a mundane wolf-like creature on the ancient shores of Pakistan, and today there are no whales. Shift the winds one way or another across northern Africa six million years ago, and humans evolve or do not evolve, as forests turn to grasslands, or not. The contingencies are endless and mind-boggling, an infinite kaleidoscope of things and events interacting with one another in ways that we may never fully understand.

Much of our amazingly improbable history has been lost to the ceaseless recycling of our restless planet. To discover the clues that remain, we can venture outward and explore the Earth's barren places, where humans and rocks converge. Or, we can look inward. Our bones tell the story of how they came to be. And each of us, within every cell, harbors a molecular library that records our ancestor's tales. Occasionally, a piece of the story comes to us in the most unexpected of ways: in the flesh and the blood of an ancient creature once thought lost.

On the day before Christmas Eve 1938, the young curator of a small museum in East London, South Africa, took a taxi down to the wharf to see what the local fishing trawler had hauled in and to wish her friends on the crew a Happy Christmas. She routinely met the incoming fishing boats to prospect for interesting specimens for her taxidermied fish collection. While rummaging through the catch, with her forearms buried deep in fish, Marjorie Courtenay-Latimer saw something new to her: "I picked away at the layers of slime to reveal the most beautiful fish I had ever seen. It was five feet long, a pale mauvy blue, with faint flecks of whitish spots. It had an iridescent silver-blue-green sheen all over. It was covered in hard scales, and it had four limb-like fins, and a strange puppy dog tail." She had no idea what it was. Nor did the crew. The Scottish trawlerman standing next to her declared that he had never seen anything like it in his thirty years of dragging fish from the sea.

Courtenay-Latimer was stumped, so she wrote to friend and mentor Professor J. L. B. Smith, at Rhodes University, about 100 miles down the coast, asking for help. Smith, at the time, was another 250 miles away in the seaside town of Knysna. The letter took almost two weeks to reach him. Upon receipt, her description and the crude sketch she included provoked an excited response from Professor Smith.

"I cannot hazard even a guess at the fish at present," he wrote, "but at the very earliest opportunity, I am coming to see it. From your drawing and description, the fish resembles forms which have been extinct for many a long year, but I am very anxious to see it before committing myself. It would be very remarkable should it prove to be some close connection with the prehistoric. Meanwhile, guard it very carefully, and don't risk sending it away. I feel it must be of great scientific value."[22]

After a rough road trip, during which Smith and his wife were waylaid midway for a week by heavy rains, they arrived in East London. The professor's instincts were correct. In his memoir, he would later recall the moment he laid eyes on the fish: "Although I had come prepared, that first sight hit me like a white-hot blast and made me feel shaky and queer, my body tingled. I stood as if stricken to stone. Yes, there was not a shadow of doubt, scale by scale, bone by bone, fin by fin, it was a true Coelacanth."

But how could this be? Coelacanths were supposed to be extinct—gone with the dinosaurs, it was thought, sixty-six million years ago.

Courtenay-Latimer had discovered the impossible. The news made worldwide headlines, and both Courtenay-Latimer and Smith were international celebrities for a time. The discovery of a living coelacanth—a "dinosaur fish," as some called it—was celebrated in newspapers, magazines, and in a Movietone newsreel.

Occasionally a species thought to be extinct, one known only from the fossil record, will turn up alive. The "resurrected" creatures are known as "Lazarus species," after the story in the New Testament book of John. These animals were, of course, never extinct. They simply had not yet been discovered alive by scientists. How many more Lazarus species are out there? No one knows.

Like the penguins, the coelacanths hold another surprise. They belong to a group of animals known as the sarcopterygians, or lobed-finned fish. They differ from other fish in that their fleshy fins are supported internally by bony, limb-like structures that extend out from their bodies. A look at the sarcopterygian branch will show coelacanths on one side and, on the other side, fish that evolved shoulders and hips and eventually proper limbs, with fingers and toes. These sarcopterygians are known as tetrapods, meaning "four feet," and include all amphibians, reptiles, dinosaurs, and mammals. Tyrannosaurs, hamsters, great horned owls, box turtles, camels, geckos, tree frogs, kangaroos, bats, Komodo dragons, coelacanths—and you—all fall within the sarcopterygian clade. All creatures with four bony limbs, including those that have lost limbs, such as snakes and whales, are sarcopterygians. We are literally fish out of water.

Now step back and take a broad view of the tree of life, and you'll see that we are members of many nested clades. We are humans, and apes, and primates, and mammals, and reptiles, and amphibians, and fish. Each of us, a menagerie. Each of us, a walking museum of natural history. Our DNA, like the city of Rome, was built and rebuilt by countless forebearers, some known, some forgotten. A shovelful of sand, a single blow with a hammer, a passageway moved a bit to the left; over time, the changes add up. Within each of our ancestral groups, our membership is complete. These are binary distinctions. You're in or you're out. There are no half fish. There are no half apes. We are apes. That's pretty obvious. But we are also fish—granted, highly derived fish, but fish none the less.[23] And so are dinosaurs.

During 92 percent of our evolutionary history, the ancestors of humans and the ancestors of dinosaurs were the same. Born of bacteria in the Archean[24] ocean, 3.8 billion years ago, our lineage is shared with dinosaurs, and with all other animals, throughout most of Earth's history. The future existence of all backboned animals teetered on a knife-edge when our tiny, wormy chordate ancestors managed to survive in the Cambrian seas of a half billion years ago. When our lobe-finned fish ancestors stood their ground along the tangled mangroves of the Devonian[25] shore, 390 million years ago, the fate of all tetrapods rested on their newly evolved shoulders. It was in the steamy forests of the Carboniferous Period,[26] about 320 million years ago, that the path of our ancestors and dinosaur ancestors finally diverged. When two distinctive groups, the sauropsids and the synapsids, emerged from our common reptile-like stock, our one road through time was irrevocably torn asunder. Evolution is a one-way street. Once parted, avenues can never be rejoined.

It would be another 89 million years before a single species of sauropsid would evolve into the first dinosaurs; 21 million years later, synapsids would produce the first mammals.[27] Each of these groups would conquer the planet—the sauropsids first, until the tables were turned by an asteroid impact.

The sauropsid-synapsid split, 320 million years ago, was one of those pivotal moments upon which all else depends. The spectacle of watching the divergence of the line that would, on one side, give rise to all reptiles, dinosaurs, and birds, and on the other, all mammals—including, eventually, a technological, spacefaring species of ape—must have been

Evolution happens so incrementally that the beginnings of momentous transformations appear mundane. The events from which massive, history-shaping changes emerge often begin as tweaks so tiny that they're barely perceptible.

Take us, for example. As different as we humans are from dinosaurs, we share a common origin. Some 320 million years ago, in the humid forests of the Carboniferous Period, a group of reptiles wandered away from their kin and started breeding separately.

At first, the two groups of evolving reptiles barely looked different. But given time—mind-bending amounts of geological time—one group would spawn the sauropsids and eventually dinosaurs. Much later, the other group would cleave again into the synapsids, which eventually led to you.

Imagine a world in which one side failed, or in which the split never occurred. Earth history is an endless collection of contingencies just like this.

an incredible sight. It wasn't. In reality, the event would have been barely noticeable.

If you could travel back in time to the Carboniferous forest of 320 million years ago, to witness the cleaving of the last common ancestral species of humans and dinosaurs, you would see an unobtrusive population of small reptile-like creatures separated into two groups by some physical barrier—perhaps a river, a canyon, or a mountain range. Cut off from breeding with their stranded kin, genetic differences would gradually accumulate, and each population would start down its own, slightly different evolutionary pathway. The success or failure of these two easy-to-overlook and hard-to-discern species would largely determine the fate of life on Earth. If one population were wiped out, the world would never know crocodiles, lizards, pterosaurs, mosasaurs, turtles, or any reptile; nor would it know tyrannosaurs, brachiosaurs, pachycephalosaurs, great blue herons, spotted owls, or any other dinosaur. If the other population failed, the world would never play host to giant ground sloths, grizzly bears, naked mole rats, hoary bats, flying squirrels, manatees, blue whales, apes, or any other mammal. And if the mammal side failed, there would be no Great Wall of China, no Acropolis, no Aristotle, no Darwin, no pottery, no disco balls, no space shuttle, no chocolate ice cream, and no you.

Earth history is filled with momentous instances cloaked in mundane garb. The next species to dominate a future age, after the hegemony of humans expires, might be living among us. Is it the humble mouse? The backyard chicken? A tide pool starfish? The inextinguishable cockroach (a sci-fi favorite)? If it were, could we recognize a species as such? Certainly not. Evolution has no trajectory and possesses no momentum. What's obvious in retrospection is often invisible in prospection. Like a tropical typhoon, propagated from a low-pressure cell, swept into being by a single flap of a butterfly wing, the events set into motion by the sauropsid-synapsid divergence were initially inconspicuous but would forever change the world.

# 4 Fossils Underfoot

When I am the first person to gaze down upon a page of Earth history, it is an abiding thrill that is never diminished by time or repetition. It's Mrs. Osler's shoe boxes full of fossils all over again but with the profundity that only an adult can feel. It's a sensuous experience, particularly in the midst of an expedition, when your body becomes weather beaten and dirty, and your hands become calloused and cracked and hard. You become a bit more like the land and a bit more connected to it. It feels primeval, forging with hand tools a deep connection to life and land and all that has ever been. Surely our forebearers, who lived in great intimacy with nature, must have reveled in the sublime pleasure of finding dinosaur bones and freeing them from their earthly confines. Surely they must have had an eye for such things, unmatched by today's urbanites and suburbanites and even those among us who still live on farms but seek shelter at night in our heated and cooled and electrified homes. Surely our rustic ancestors, whose senses were tuned to appreciate the subtle advertisements and ominous warnings of nature, must have had a keen eye for seeing and appreciating fossils. This seems as though it must be true, but it is not. The eyes collect light, but it is the prepared mind that sees and creates story. And a fossil without a story is just a rock.

Prior to the turn of the nineteenth century, there was no coherent Earth story that incorporated the fossil record even though, in the right geological circumstances, encounters with fossils must have occurred frequently for people living in agrarian societies or among nomadic tribes. Farmers working the soil with only their draft animals and their hands experienced a familiarity with the land that few of us can imagine today. Fossil sea creatures, mammoth tusks, and dinosaur bones would not have gone unnoticed by the ploughman, the goatherd, the reaper, and so many other preindustrial professions that required a boots-on-the-ground, hands-in-the-dirt intimacy with rocks. The surveyor and the drainer, the catchpole (tax collector) and the wandering merchant, the woodcutter and the forester, and, of course, the miner and the stonecutter must have all encountered fossils in the course of their work. Without a scientific framework to contextualize these observations, though, chance encounters with our prehistoric past may have been simply ignored, or viewed as the curious, spontaneous emanations of the Earth, or rolled into vernacular folk legend. Before science existed to draw out story from bone, concocted tales turned some fossils into folklore.

In her illuminating book *The First Fossil Hunters,* Adrienne Mayor argues that even though "it may be true that no natural philosopher—not even Aristotle—articulated a formal theory to explain fossils," allusions to the bones of monsters and giants in ancient texts "are evidence for a native natural history."[28] For example, Mayor contends that fossil remains of ceratopsian dinosaurs from the Gobi Desert may have inspired the myth of the griffin.

Legend holds that griffins possessed the body of a lion and the wings and the head of an eagle (sauropsids and synapsids reunited at last) and were jealous guardians of caches of gold. It would be easy, and possibly correct, to dismiss the lore of the griffin as a completely fictional construct. Mayor argues, though, that fantasy sometimes germinates from a kernel of reality; that legendary monsters sometimes spawn from the folk paleontology of ancient peoples.

Moreover, a number of other ancient writers appear to have grasped that fossils represent preserved organisms from a previous age. In *The Seashell on the Mountaintop*,[29] Alan Cutler writes, "The very earliest of the Greek philosophers, the so-called pre-Socratics, made it a keystone of their various theories of the world six centuries before Christ." For some ancient Greeks and Romans, the meaning of fossils as representations of earlier life-forms was intuitive, as it is for us today. It's common sense that a clam once lived inside a clamshell and that a leg once surrounded a leg bone. Following the fall of Rome, though, this understanding, like so much else that was once understood, was lost for a millennium.

Throughout the Middle Ages and much of the Renaissance, the best-educated people in Europe rejected the seemingly obvious biological origin of fossils. Spontaneous generation was considered a fundamental natural process, responsible for maggots on meat, fruit flies on produce, and curiously shaped objects in rocks that superficially and coincidentally resembled animals. With your contemporary eyes, even if you've never paid particular attention to fossils, you would immediately recognize a trilobite, or an ammonite, or a brachiopod as a fossil. Maybe not by name, but you would recognize the shapes and patterns and textures of life. You would recognize these stony time capsules from a bygone age as biological in origin.

Fossil shark teeth, in particular, are some of the easiest remains to identify as fossils, although identifying the particular type can be tricky. Fossil shark teeth look exactly like, well, shark teeth. In many cases, even the tooth enamel remains, making them shiny, although usually black, because of minerals taken up from groundwater into the tooth matrix. (Or, as I joke with the schoolkids visiting our fossil park, because they haven't brushed their teeth in sixty-six million years. "Let that be a lesson to you," I usually add.) I would wager, with confidence, that if I presented you with a fossil shark tooth, you would immediately identify it as such. It would be no more difficult for you than identifying an apple. Even if you had no experience with apples—even if you were a member of an uncontacted Amazonian tribe, let's say—you would undoubtedly recognize an apple as a fruit of some kind, though you would not have a unique word for the exotic pome. But if I were presenting a fossil tooth to you, and you were a seventeenth-century inhabitant of Europe, I would make no such bet. In fact, I would wager just the opposite. There would be almost no chance that the tooth's ancient origin would be apparent to you. It couldn't be. You would not have believed the Earth to be old, and you would have had no conception of life beyond than that which was apparent in your present. In short, you would have believed the Earth to be no older than about six thousand years, and you would have subscribed to a mash-up of Old Testament creationism and the Aristotelian concept of fixity of species, which holds that species are immutable over time.

To the inhabitants of Malta, who lived on and cultivated ancient marine deposits, fossil shark teeth were common occurrences in their landscape. Farmers working the striking blue clay marls of the Mediterranean island would frequently turn up the teeth of fossil sand sharks and mackerel sharks. Those who cultivated the creamy yellow beds of limestone would encounter the monstrous teeth of *Carcharodon megalodon*, the record-setting megapredator of the Miocene seas.

They did not recognize the fossils for what they were, though. These shark teeth were never in the mouth of a shark. They called them glossopetrae: tongue stones, representative emanations of the mystical world. Elsewhere it was thought that glossopetrae fell from the sky. The Roman naturalist Pliny the Elder posited that this petrifacted precipitation occurred only on black, moonless nights, which explained, of course, why no one had ever actually seen tongue stones drizzling from the sky. On Malta, they knew better. They dug tongue stones from the ground and sold them as a powerful talisman, which bestowed upon the bearer protection against poison. Legend held that while returning to Rome, the Apostle Paul was shipwrecked on the island. While setting a cooking fire, a vicious serpent leapt from the firewood and struck the disciple in the arm. Unscathed, Paul dashed the serpent to the ground and cursed it, forever more depriving it and its kin of their deadly venom. The Maltese believed that periodically nature would commemorate Paul's miracle with the production of glossopetrae resembling the fangs of a serpent, thus explaining their reputed prophylactic effect against poison.

Throughout most of postclassical history, the greatest minds in philosophy, science, and Western theology all regarded the Earth and its life to be the relatively recent creation of a divine architect. Open an eighteenth-century King James Bible and turn to Genesis 1:1—"In the beginning God created the heavens and the earth"—and you are likely to see the number 4,004 annotated in the margin. That's 4,004 BC, the supposed year of creation. This date comes from an exacting piece of scholarship performed by the Irish theologian James Ussher, archbishop of Armagh and primate of All Ireland. (This title refers to his primacy over the church of Ireland and not to his simian ancestry, as I'm inclined to read it.) Using multiple sources, Ussher worked back from the death of the Babylonian king Nebuchadnezzar, through the reigns of other Old Testament kings, and through an unbroken series of "begats" recorded in the Bible, from Solomon to Adam, to arrive at his date. Through further supposition, he narrowed the date to October 22, 4004 BC—a Saturday, around sundown.

Precision and accuracy are two very different qualities. Although Ussher worked with great precision, his conclusion was monumentally inaccurate—precisely wrong. The magnitude of his error is the equivalent of looking at the Empire State Building, which is 1,454 feet tall, and believing it to be less than 1/32nd of an inch tall, or less than the thickness of two dimes. Still, it was a credible attempt, and he was not alone in his gross misapprehension of the antiquity of our planet. Scientific luminaries no less bright than Isaac Newton subscribed to a young-earth Mosaic account of creation. Charles Darwin himself was a young-earth creationist when he first strode up the gangplank of the HMS *Beagle* in preparation for its five-year voyage around the planet.

On a young planet, fossils make no sense. There is no ancient past to record. If you believe the Empire State Building to be no higher than a couple of dimes, your concept of the origin and function of the Empire State Building will be profoundly altered, and profoundly flawed. It seems impossible to us to see a fossil clamshell or shark tooth as anything other than biological in origin. But if we begin with the enfeebling assumption of a six-thousand-year-old Earth, the truth of it will be forever out of reach. On a young planet, species must be divinely created and fixed as they are today. There is no time for evolution, and extinction cannot be allowed. How could it be? If creation were a once-in-history divine act and extinction operated naturally, picking off species over time, the Earth would have no way to replace lost species and would eventually run out!

Thus, without the framework of deep time, fossils went generally unnoticed or were viewed as oddities of nature. No doubt, humans living in the desolate and craggy places of the world must have occasionally stumbled upon the odd dinosaur bone. Most such remains, having been mineralized—turned to stone—were probably not recognized as bone. Certainly none was recognized as dinosaur bones. There was not even a word for dinosaur.

An eighteenth-century shepherd might have regarded an ammonite fossil in the same way that we might pause for a curious moment to examine a demonic visage sculpted by the random wrinkles of the bark of an oak, or a vaporous cloud elephant wafting lazily overhead, or a cliff face crumbled in just the right way to reveal the jagged figure of a weather-beaten hero. Humans are pattern-recognition machines. We see pattern everywhere, real and imagined. If you're not inclined to believe in tree spirits, though, a face in a tree is an amusing thing to note but nothing of consequence. Picking out animal figures in the clouds is a child's game that no one takes seriously. If you're not aware of the antiquity of Earth, then ammonites, trilobites, dinosaur bones, and other fossils are interesting rock formations, but, like a menagerie in the sky, illusionary.

"I wouldn't have seen it if I hadn't believed it," read the sign on my graduate advisor's door. Put another way, seeing is believing, but believing is also seeing. Or in more scientific parlance, observing fosters awareness, and awareness prepares the mind to observe. Field scientists call this a search image. Whenever I enter a field area for the first time, I feel blind to the fossils underfoot. Fossils of the same type can appear quite differently from one place to another, because of local geological conditions. They might be crumbly and glistening with a frosting of gypsum in the Egyptian Sahara. They might be rock-solid and painted dark with a patina of iron in the rugged outcrops of Patagonia. They might be waterlogged and olive-green, stained by ancient marine sediments, in the bountiful marl pits of New Jersey. Some bones turn to opal, malachite, or pyrite. The possibilities are endless. Because of this, I find that it takes me about three days to really get my "eyes" at a new site, to really establish my search image. After this, it's like having fossil radar. Fossils that I would have stepped over two days before become as apparent as the brightest stars on a cloudless night. Getting to this point, though, requires active observing and a subconscious awareness that fossils are a thing in this world.

There is no better place to develop an eye for fossils than in the bountiful badlands of southern Patagonia. There I spent five frigid austral summers digging up the remains of the giant herbivorous dinosaur *Dreadnoughtus schrani*. During one field season, we were joined by an archaeology student from the University of Buenos Aires, Brenda Gilio. An expert in the early human occupation of southernmost South America, she decided to take a few days' holiday from her archeological site, about sixty miles to the south, to help us with our titanic task of getting *Dreadnoughtus* out of the ground. Despite widespread confusion among the laity, paleontology and archeology are widely separated disciplines with almost no overlap. (Please stop asking archeologists about dinosaurs; it really ticks them off. And if you ask me about pyramids, you'll find that my knowledge is limited to what I read in *National Geographic*.)

The morning after her first night in our camp, I accompanied Brenda on the ten-minute walk up a rocky arroyo to our quarry. This was our third field season there, and by this time, I had walked that walk hundreds of times. Brenda had no eye for fossil bone, and I had to point out the fragments that lay scattered in the dry streambed that was our path. Then she stopped and picked up a rock. She showed it to me, and I gazed vacantly upon it. Yep, you found a rock, I thought. This might be a long three days. Then she smiled the wry smile of someone enjoying a secret and told me that it was a Native American hand ax. Instantly, the teardrop shape of the stone became apparent to me, and the many handcrafted facets became wondrously obvious. By transferring one simple thought from her brain into mine, she had magically and instantly caused an unremarkable chert cobble to transform into a precious artifact. How could I not have seen that before? Because I have no search image for such things—I'm not an archeologist. Paleontologists tend to ignore recent unconsolidated deposits. We look through them and focus on the ancient rocks that lie below. Where I saw background noise, Brenda saw signal. Before we arrived at the quarry, she had found two more hand axes. After she left, we never found another. With only a layman's awareness of such objects, my mind is not adequately prepared to see them. And, apparently, I don't.

Prior to the nineteenth century, no one had a search image for dinosaur bones. Without minds prepared to observe their occurrence, they went largely unnoticed; anonymous rocks in the landscape. It's hard to imagine the number of bones that must have been turned under by the ox man's indifferent plough. What priceless fossils have been neatly stacked into stone walls, to conclude their long existence with an unlikely chapter battling the forces of weather, insensate livestock, and feuding neighbors? How many jaws, teeth, ribs, and vertebrae from the ages before humans have been trod upon, unnoticed, by pilgrims, refugees, adventurers, soldiers, and merchants?

Dinosaurs, themselves, though they were numerous and speciose, were nothing more than a tiny twig on the cornucopian tree of life, which first sprouted from the abiotic Earth at least 3.8 billion years ago. At every point during life's long tenure, save the first little bit, extant creatures have represented but a slender fraction of all that has ever been. Prior to the advent of paleontology and geology, we were utterly blind to this past; blind to deep time. And prior to the age of exploration, even our present lay largely obscured beyond our parochial horizons. So borne upon our little perch at the end of our sprig of primates, we knew little of the tree's canopy and nothing of the timber on which it rests.

The wall of ignorance that separated us from an awareness of 99.9 percent of all the creatures that have inhabited our planet was slow to crumble. The first few cracks formed when the true nature of glossopetrae came to light. In a 1554 compendium on Mediterranean fishes, French physician Guillaume Rondelet suggested that tongue stones were, in reality, the petrified teeth of sharks. His argument, though, did not seem to persuade. The mystical powers of the glossopetrae would not be so easily dissipated. Sixty-two years later, a Neapolitan lawyer named Fabio Colonna launched his own investigation into the nature of glossopetrae. After rigorously examining many tongue stones inside and out, he wrote, in 1616, "nobody is so stupid that he will not affirm at once at the first insight that the teeth are of the nature of bones, not stones." Still, the people were not convinced. Neither Rondelet nor Colonna presented a mechanism for the petrification of teeth from actual sharks. How could shark teeth end up on land? That idea must have sounded silly to most. How could they turn to stone? What about the curative powers of

All of Earth's past comes to us in the form of rocks and the fossils they bear. For millennia we were blind to the vast histories contained in stone, blind to the fantastic creatures trapped in rock, just beneath our feet. The discoveries of deep time and the biological origin of fossils opened our eyes to billions of years of events that preceded us.

glossopetrae? Surely the legends better explained that. And there were already plausible explanations for tongue stones. The Apostle Paul certainly could have been marooned on Malta, and who knows what miracles he could command? So why not serpent fangs? Why not woodpecker tongues, as thought by some? Trees, after all, are pretty hard. Why not spontaneous generation? This was known to occur. All these explanations would have sounded more feasible to a seventeenth-century ear than sharks on land!

Another half century passed before a Danish anatomist dissected the head of a great white shark that had turned up on the Tuscan coast near Livorno, Italy. Niels Stensen, turned Nicolas Steno, was living in Florence as a guest of Ferdinando II de' Medici, the grand duke of Tuscany. The twenty-eight-year-old Steno was celebrated for his anatomical insights and his deft touch with a scalpel. He was also familiar with glossopetrae and probably knew of the hypotheses of Rondelet and Colonna. During his examination of the great white shark, it quickly became clear to him that tongue stones were actually petrified shark teeth. In fact, according to some taxonomists, great white sharks arose from megalodon. So if Steno had a megalodon glossopetra from Malta in his hand for comparison, the resemblance must have been striking. What was different this time, according to Steno biographer Alan Cutler, was that "it was also clear to Steno that the tongue-stone question was really only a special case of the general problem of fossil seashells and other 'marine bodies' dug from the Earth in places far from the sea." Steno realized that fossils had to be studied within the context of the rocks bearing them. And thus he began systematic geological study, though he did not have that term. In the course of his investigations, he deduced four principles that described the manner in which sedimentary rocks are laid down, accumulate as strata, and are crosscut by geological processes. These ideas today are known as Steno's laws and are a staple of university Geology 101 courses.

Steno later turned his attention to Catholicism and then asceticism, ultimately dying from malnutrition in 1686 at age forty-eight. One wonders what progress he would have made had he dedicated a full-measure life to science. Steno's laws would later form the basis of the subdiscipline of geology called stratigraphy, the study of rock strata, which is particularly important in mining and petroleum exploration. And his theory for the biological origin of fossils slowly gained traction. By the early eighteenth century, fossils were generally considered biogenic (created by organisms), but their origin, along with Steno's stratigraphic observations, were rolled into Noachian flood mythology. Young-earth creationism prevailed. The volumes of Earth history were now known to exist—progress—but they remained, except in a superficial way, indecipherable to humans.

Unless the true nature of time could be apprehended, humans would never grow to understand the richness, drama, and vicissitudes of Earth history. Who are we? How did we get here? What is our place on this planet? Answers to these profound questions lay just beneath our feet, but we were blocked from seeing them by our faulty assumptions. Blinkered by young-earth dogma. A breakthrough was needed to reveal a fundamental truth about our planet, a truth that would allow all the rest to fall into place—the Earth is old.

# 5 Deep Time

Time, deep time, is the quintessence of geology, the thing that makes rocks and fossils make sense. Our senses are not well tuned to perceive the full panoply of natural phenomena. They evolved to help us deal with the here and now. Threats, food, and mates must be attended to with urgency. We live in the now, and living memory is brief. Our lives play out in only tens of years. It's a mournful fact. Even the historical record is basically now. The years 2017 and 1066 are essentially the same moment when viewed over the sweeping vastness of Earth history. Draw the divisions of geological time to scale, from beginning to end, on a sheet of paper, and the entire human experience falls within the breadth of your last pencil stroke. Geologically, the human diaspora from Africa, the settling of the Fertile Crescent, the classical period, the industrial revolution, the space age—it's all now.

Deep time is unfathomable. Its eons span billions of years, its eras, hundreds of millions. The Cretaceous Period lasted for nearly eighty million years yet represents less than 2 percent of Earth history. Grappling with deep time is one of the epic struggles in the ascendancy of humanity. Wresting it from the rocks and carving it into comprehensible bits is the $E=mc^2$ of natural history. The geological time scale looks so simple—a bunch of rectangles, nested in rectangles—but its meaning is a towering achievement of science, and it was extracted from the Earth at great cost.

By themselves, the products of Earth tell no tales, word salad on the landscape. The wondrous stories preserved in the annals of Earth history were first revealed to us only after the discovery of deep time. Without it, there's rock collecting and there's fossil hunting, but there is no geology, there is no paleontology, there is no science to be done. With all but the most recent paragraph of Earth's story written prior to the emergence of *Homo sapiens*, the pages have always been there for us to read, waiting silently underground for our graduation into a geologically literate species. The magnitude of error inherent in young-earth creationism had long prevented any coherent reading of the rock record. Blinded by mysticism and by our cognitive bias toward the present, we sat, until quite recently, on our tiny sprig, in complete ignorance of the tree below. Finally, near the turn of the nineteenth century, a man whose name you may never have heard penned these words: "The Earth reveals no vestige of a beginning, no prospect of an end."[30] It turns out that our historical past had a past, and *that* past had a past, and so on. Suddenly the world appeared to be a much different place and our place on it, much smaller.

With those words, and the two volumes[31] that would follow, James Hutton gave us deep time. A man of the Enlightenment, a physician turned experimental farmer, Hutton spent years traversing the craggy mountaintops, heather-carpeted moorlands, and rocky coasts of Scotland in obsessive pursuit of the simplest of questions: What made the mountains? What made the cliffs? What made the rocks? What made the soil? In a top-notch piece of geological sleuthing, building on Steno's four laws, he deduced correctly the complex series of events that led to the deposition, crosscutting, and uplift of the sandstone beds of Siccar Point on the North Sea coast.[32] Hutton's brilliant insight was that this impressive geological edifice and, moreover, the entire landscape, was the work of ordinary, everyday, slow-acting processes: sand grains saltating in the breeze, the swash and backswash of waves, the ceaseless trickle of the babbling brook, the timeless drumming of raindrops. How could it be, though, that these gentle processes were capable of moving mountains, filling out bottomlands, and sculpting coastlines? Time. Given enough time, the insensible lapping of the waves, the decades-long meander of the stream, the pelt-pelt-pelting of sand grains driven on the wind, and other delicate acts of natural labor were capable of doing real work. But it would take a lot of time, vast tracts of time, deep time.

There it was. The Earth was old. In his masterpiece *Theory of the Earth, with Proofs and Illustrations,*[33] published in 1795, Hutton gave agency to our planet. No longer the passive product of a heavenly actor, it was earthly processes that created our world and continue to reshape it today through mundane, slow-acting, and—this is key—observable phenomena. This was no legend, passed from

The world that we see today is only the latest version of our planet. Many past worlds have been our Earth. Each one leaves behind layers and clues for us to discover: a fossilized dinosaur femur here, the rocks of an ancient seafloor exposed on a mountaintop there. Each one fits within a story more grand than the mind can fathom, a story that stretches back beyond humans, beyond dinosaurs, beyond continents, even beyond life itself.

Only when we grasp the astonishing
scale of deep time, revealed to
us in the vestiges of countless
antecedent worlds, do the clues
reveal their meaning.

the mouths of countless illiterate forebearers. This was no tale of a divine superhuman in space. This was a creator that could be reckoned with, studied, and deconstructed in the crucible of science.

A common thread among origin myths is that the processes of creation are not the processes of today. The house is built. The house is lived in. Those are two different sets of processes. Living in the house tells you little of how the house was built. If you have an eye for such things, you could certainly discern some of the methods of construction, such as two-by-four framing verses post-and-beam construction. But lost would be the identities and the stories of the builders themselves. Were there arguments, injuries, cost overruns, or shortages? What was the season? How was the weather? Were the builders on schedule? Was the project profitable? Unless you participated in the construction of your home or (less likely) your apartment building, you probably cannot answer these questions for your current domicile. The link is broken.

With no connection to the past, observing nature may be interesting, but it has no bearing on the magical moments of inception, during which the firmament was woven together to create our world. This discontinuity makes sense if the Earth is young. If the geological record is really less than six thousand years old, there is no time for slow-acting processes to affect much change. Look around you. The mountains, the oceans, the rivers, the forests, and the animals must have all come into being rather suddenly in a young world. There is no time for other scenarios to operate.

Imagine that you truly believed the Empire State Building to be no more than two dimes high. Now imagine your mental shock upon the revelation that it is actually more than a quarter mile high. The canvas upon which Earth history was painted was not a postage stamp, it was a football field! Everything didn't have to happen at once. Six days of divine creation turned into eons of mindless sculpting, the unconscious world crafting itself. The creator, it turns out, was all around us, visible each day to us: the stream, the wind, the waves, the ice and rain, and much later, we would learn, the slowly roiling mantle beneath the crust and the imperceptible crunching, stretching, and sliding of tectonic plates.

Like so many scientific revolutions, Hutton's work made little impact initially, aside from drawing fire from a number of prominent detractors. His *Theory of the Earth* was a rambling double volume, which required readers to plough through opaque and convoluted passages in order to reap the bounty of his revelations on the true nature of Earth and time. Gravely ill while he wrote it, Hutton lacked the vigor and time to personally advocate for his magnum opus. Adherents of the catastrophist views were quick to attack his work, and they did so relentlessly and effectively. Following his death in 1797, his friend and mathematician John Playfair mounted a spirited and loving defense of Hutton's theory,[34] but by

1813, the supporters of young-earth ideology had authored convincing refutations[35] of Hutton's thesis.

Only eight months after Hutton died, a baby was born to Frances and Charles Lyell, an English family living in Scotland, not far from the pretty little Glen Tilt. Hutton had visited this valley, which, with its intrusive veins of granite, surrounded by sedimentary rock, had given him his best evidence that heat from below stoked uplifting forces that held the power to rejuvenate the Earth's crust. The child born in this illustrative terrain would develop a supple mind and finely honed skills of observation. Yet he would learn from prominent men that Hutton's works were nonsense and would accept on authority the catastrophist view.

Charles Lyell, son of his namesake, grew up in Scotland and in England and attended Oxford University, ostensibly to study law. His thoughts, though, were drawn inexorably to natural things, and geology soon became his passion. He knew of James Hutton, but mostly as an object of derision. His Oxford geology professor, William Buckland, the first academic geologist in England, was an ardent catastrophist. Buckland believed the Noachian flood to be the latest in a series of calamities. To him, it was quite obvious: "A universal deluge at no very remote period is proved on ground so decisive and incontrovertible, that had we never heard of such an event from Scripture, or on any other authority, Geology of itself must have called in assistance of some such catastrophe." Indoctrinated by the weighty pronouncements of his esteemed mentor, the young Lyell was brought into the fold of catastrophist philosophy.

In 1819 Lyell graduated from Oxford and moved to London to study law, but again the rocks tugged at him. He soon began traveling across Europe to see for himself what he had read about in recent scientific reports. Back in England, he visited surgeon and avid fossil collector Gideon Mantell, who had been finding curiously large bones in a quarry in the town of Cuckfield, about forty miles south of London. Lyell visited the Isle of Wight and saw chalky marine strata perched high over the sea, evidence of uplift. He traveled to Paris and marveled at interfingering layers of freshwater deposits and saltwater deposits, which appeared to have occurred not by sudden disruptions but by the gentle alternations of environments. Catastrophism could not explain this, but Hutton's cycles could. This, perhaps, drove him back to Scotland, where he observed freshwater limestone in the process of forming. Catastrophists claimed that all limestone was from a bygone age—the thread between past and present processes broken.

Then Lyell paid a visit to Hutton's old friend and confidant James Hall, a University of Edinburgh professor and one of the last living supporters of the nearly extinguished Huttonian view. Hall took Lyell by boat—just as Hutton had taken him three and a half decades before—to visit Siccar

Point. Seeing the outcrop must have shaken Lyell's confidence in the catastrophist philosophy in which he'd been so thoroughly indoctrinated. Like anyone who leaves a cult, discovers a dark truth about a relationship, or sees a hero exposed as a villain, Lyell was presented with confounding and troubling evidence that required a 180-degree reversal of his world view; a reversal of the ideas on which his identity was pinned.

These were propitious moments in humankind's discovery of the Earth's ancient past. While Lyell was flirting with Hutton's deep-time heresy, an awareness was emerging in the minds of a few English naturalists—giant extinct beasts had once roamed our planet. Without realizing it, we had begun to discover Earth's dinosaurs. As if a fog was lifting to reveal what was always there, the view was murky at first.

# 6 Thunderclaps

The first dinosaur specimens to undergo scientific study didn't cause much of a stir. Naturalists were flummoxed by the size and strangeness of dinosaur anatomy, and the public was largely unaware that ancient monsters were turning up in the woodlands of England.

In 1822 British physician Gideon Mantell and his wife, Mary Ann, made a small collection of large fossil teeth, which they found in the Tilgate Forest, near Cuckfield, sixty miles south of London.[36] Gideon, a fine naturalist, did not know what to make of them, recording in his journal that he was "not in the hope of being able to elucidate their nature." Various English authorities pronounced them to be the teeth of a fish or a mammal. Mantell's friend Charles Lyell took several teeth with him to France. At a Parisian soirée, he showed the specimens to the great French naturalist George Cuvier, who declared them to be rhinoceros. Mantell, who thought the teeth looked reptilian, must have been crestfallen to receive Lyell's note containing Cuvier's contradicting and authoritative diagnosis. While most would have considered this definitive, Mantell was not convinced. He suspected that the teeth were related to the large bones he was finding—bones he thought to be saurian in nature, a Linnaean term that, to him, would have meant lizards and crocodiles. (Today the term is seldom used, and includes lizards, snakes, and archosaurs such as crocodiles, pterosaurs, and dinosaurs.)

Meanwhile, William Buckland had been acquiring some large bones of his own from quarry workers in Stonesfield, twelve miles from Oxford. The remains were scant and provided no clear picture: a bit of lower jaw with a single tooth, two vertebrae, a partial pelvis, two ribs, a thighbone, and a scrap of toe, all from various individuals. Buckland believed he had pieced together the bones of an ancient giant reptile. On February 20, 1824, he presented his results to the Geological Society of London. He named the monster *Megalosaurus*, meaning giant lizard, for that is what he believed it to be.

This, perhaps, inspired Mantell to send his mysterious teeth back to Cuvier, asking for a reexamination. This time Cuvier agreed with Mantell: the teeth, he wrote, were of saurian origin. Emboldened, Mantell announced his conclusions at the venerable Royal Society of London, nearly a year to the day after Buckland's unveiling of *Megalosaurus*. Mantell named his ancient lizard *Iguanodon*, whose teeth, he thought, resembled those of living iguanas.

At this point, the existence of dinosaurs was still unknown and would remain shrouded in the mist of the past for nearly another two decades. Buckland's and Mantell's giant lizards were, in fact, dinosaurs, and the pair of fossil collectors had, without realizing it, authored descriptions of the first two dinosaurs given scientific names.

Six months after Mantell published his paper[37] about the *Iguanodon*, the young Charles Darwin matriculated into the University of Edinburgh, where he enrolled in a geology course taught by Hutton's old nemesis Robert Jameson. There Darwin was inculcated with a full dose of unreconstructed diluvian, young-earth geology. Jameson even brought Darwin to see some of Hutton's most cherished outcrops, deriding the deceased geologist on his own turf. Convinced, Darwin attended regular meetings of the Wernerian Society, dedicated to the study of the catastrophist worldview.

Meanwhile, Lyell traveled to postrevolutionary southern France, where he saw alternating layers of volcanic basalt and river gravel, clear evidence of both uplift and cycles. Then he headed south to Italy, where his conversion became complete. On the island of Ischia, off the coast of Naples, he found dramatic examples of coastal uplift occurring within historical times: shell beds containing the same species seen today but high in a cliff. This was not supposed to happen. The processes that created mountains were supposed to have been contained to another age. Yet there it was, a thunderclap. The old stories were wrong. The geology that he had learned at the most prestigious university from the most celebrated minds in the natural sciences was wrong. The Earth was alive, Lyell realized. Geology wasn't just a thing of the past, it was happening right now. In a letter home to his sister, his excitement was palpable: "I will let the world know that the whole Isle of Isk [*sic*], as the natives call it, has risen from the sea 2,600 feet since the Mediterranean was peopled." His tour of the Mediterranean continued and reinforced what he now knew. There was a link between processes in operation today and the Earth's past. The thread remains. Earth history is a continuum. Earth history is knowable.

With the zeal of a convert, Lyell wasted no time in publishing his newfound understanding of the Earth. In July 1830 he published the first volume of *Principles of Geology, Being an Attempt to Explain the Former Changes of the Earth's Surface, by Reference to Causes Now in Operation*. In it he gave ample credit to Hutton, writing that he was the first "to explain the former changes of the Earth's crust, by reference exclusively to natural agents." Hutton's theory, Lyell wrote, "was more shocking when coupled with the doctrine that all past changes on the globe had been brought about by the slow agency of existing causes." With his meticulous observations, scrupulous logic, and elegant delivery, Lyell would prevail. Standing on Hutton's shoulders, as all geologists do, he would enjoy the victory that eluded his predecessor. Approached with an open mind, his assertions, most would eventually agree, were undeniable. Over the course of three volumes, Lyell essentially put an end to biblically inspired geological theories and forged an unbreakable link from the present to the past.

Writing about Lyell's work for the British literary and political periodical *Quarterly Review*, the philosopher and historian of science William Whewell dubbed Lyell a "uniformitarian" and his opponents "catastrophists." (Whewell, famous for his neologisms, also invented the term *scientist.*) The names stuck, and today uniformitarianism is the bedrock principle of geology and paleontology, and is often encapsulated with the phrase "The present is the key to the past."

The next summer, while Lyell was laboring on this second volume, twenty-two-year-old Charles Darwin received a letter inviting him to join the crew of the HMS *Beagle* as the ship's naturalist on its voyage around the world.[38] The young Darwin had yet to find his life's passion and was, in fact, a bit of an idler. Chided by his upstanding father, "You care for nothing but shooting birds, dogs, and rat catching, and you will be a disgrace to yourself and all your family," Darwin was, perhaps, looking for a fresh start and a chance to test his mettle.[39] Despite his father's initial objections, Darwin jumped at the chance and, with the help of an uncle, was able to win the older Darwin's acquiescence, if not his support.

In preparation for the voyage, Darwin wanted to bolster his geological skills. He joined Cambridge University geologist Adam Sedgwick as his assistant on a three-week excursion through the mountains of Wales. Sedgwick was committed to field observation, but, like others of his time, he was a catastrophist who thought the world came into being through a series of relatively recent convulsions, culminating in Noah's flood.[40] No doubt, Darwin's time with Sedgwick reinforced the young naturalist's view of a young earth shaped by divine calamity.

Prior to embarking, Darwin received a parting gift from the ship's captain, Robert FitzRoy: volume 1 of Charles Lyell's *Principles of Geology*.[41] The *Beagle* weighed anchor and departed Devonport, England, on the second day after Christmas 1831. Yuletide merrymaking had laid low a good portion of the crew and scuttled their plans to depart the day before. Twenty days later, the *Beagle* anchored off St. Jago (Santiago) in the Verde Islands, about 400 miles off the northwest coast of Africa.

Darwin was eager to explore and immediately set out on foot. He visited the small town of Porto Praya, devoured fresh oranges, and tasted a banana, which he did not like. Taking the long way back, excitement filled him as he trod for the first time in his life on an exotic landscape. "I returned to the shore," he wrote, "treading on Volcanic rocks, hearing the notes of unknown birds, & seeing new insects fluttering about still newer flowers. — It has been for me a glorious day, like giving to a blind man eyes."[42] He would spend the next three weeks exploring the island. The landscape of St. Jago is desolate, but no matter. "[T]o anyone accustomed only to an English landscape," he wrote, "the novel prospect of an utterly sterile land possesses a grandeur which more vegetation might spoil."[43] He must have felt like a real naturalist. Here was a chance to reinvent himself, to show that he wasn't the slacker that his father saw. He must have felt like a grown-up, blazing, for the first time, his own path in life.

Ask any geologist, and he or she will tell you of the singular rapturous delight of entering a land devoid of plants. Darwin, a geologist of sorts, was no different, writing that "The geology of this island is the most interesting part of its natural history."[44] In an epiphanic episode, strikingly similar to Lyell's eye-opening experience on the island of Ischia, Darwin observed "a perfectly horizontal white band in the face of a sea cliff," resting about forty-five feet about the sea. Clambering up to get a better look, he saw "numerous shells embedded, such as now exist on the neighboring coast."[45] The shell bed was bounded from above and below by layers of lava. Darwin noted that the uppermost several inches of the shell bed had been baked into solid rock by the heat of the overlying lava—*contact metamorphism*, we would call it today.

Ever a sponge for knowledge, Darwin had soaked up Lyell's volume sometime before or during his explorations of St. Jago. As he pondered the outcrop, he saw in it a history of cycles and of uplift. He saw the Earth working and reworking itself. The geology that he learned from Jameson and Sedgwick could not explain this. These sea creatures had populated a submerged shelf of volcanic rock. While still beneath the waves, a subsequent lava flow entombed the shells, baking the ones on top. Then the whole section was uplifted forty-five feet into a sea cliff.[46] Darwin saw that no miracle was required to explain this outcrop. He began to see the rocks through the eyes of Lyell. These pages of Earth history, he thought, were written by the slow agency of existing causes—existing causes and time.

The alacrity with which Darwin was able to replace his deeply held worldview, in the face of countervailing evidence, is a window into the agility and suppleness of his mind. His ability to embrace a radical, even blasphemous, interpretation of the rocks on St. Jago foreshadowed what would become his greatest quality as a scientist: a willingness to follow the evidence, wherever that might lead. As the *Beagle* departed for its crossing to South America, Darwin contemplated all he had seen and how it had changed him, writing in his diary that the memories "will never be effaced from my mind."

On July 20, 1832, the *Beagle* was off the coast of Uruguay, approaching the mouth of the Río de la Plata. At this very moment, quarry workers in the Tilgate Forest, back in England, set light to a large charge of gunpowder packed into a siltstone cliff face. As the dust cleared following the explosion, the miners noticed something odd: a jumble of bones set into the rock. Word reached Gideon Mantell, who purchased the bones, still in the rock. He recognized the remains as another large saurian; the most complete yet. But this specimen showed startling features not seen in either *Megalosaurus* or in the bones he related to *Iguanodon*. There were broad plates and fearsome sharp spikes—a giant armored lizard! Mantell named the new creature

*Hylaeosaurus armatus*, "armored forest lizard." It was, unbeknownst to him, the third named dinosaur.

In the fall of 1832, a parcel containing Lyell's just-published second volume of his *Principles of Geology* reached Darwin in Montevideo, Uruguay. No doubt, he must have dug into it with great anticipation. By the time Darwin was rounding Patagonia through the Strait of Magellan at the tip of South America, he was paraphrasing Lyell—borderline plagiarism, really. Reflecting on the creation of the strait, Darwin wrote in *Voyage of the Beagle*, "we must confess that it makes the head almost giddy to reflect on the number of years, century after century, which the tides, unaided by a heavy surf, must have required to have corroded so vast an area." He is clearly channeling this passage from *Principles of Geology*: "The imagination was first fatigued and overpowered by endeavoring to conceive the immensity of time required for the annihilation of whole continents by so insensible a process."

Darwin was now a radicalized uniformitarian. A deep-time evangelical. While traveling up the west coast of South America, he wrote to his second cousin William Darwin Fox, a clergyman and amateur naturalist: "I am become a zealous disciple of Mr Lyells views, as known in his admirable book.— Geologizing in S. America, I am tempted to carry parts to a greater extent, even than he does."[47] Many years later, upon the death of Charles Lyell, Darwin eulogized him, writing these words: "How completely he revolutionized Geology: for I can remember something of pre-Lyellian days. I never forget that almost everything which I have done in science I owe to the study of his great works."

Why did Darwin, who himself had revolutionized a field of science, place so much value on Lyell's work? It's simple. Lyell—and, by extension, Hutton—forced a paradigm shift that created the thought space in which Darwin constructed his science. *On the Origin of Species by Means of Natural Selection, or the Preservation of Favoured Races in the Struggle for Life*, Darwin's 1859 masterpiece, is set firmly in the intellectual milieu of uniformitarianism. Darwin's seminal contributions to human thought—evolution by natural selection and universal common descent—make no sense on a young planet. Tiny changes, from one generation to the next, would amount to very little on a six-thousand-year-old Earth. For in a catastrophist world, evolution is a feeble force, impotent to effect much change. Robbed of her chisel and mallet, it could not have been nature who sculpted the biosphere. And with species fixed from a burst of creation, descent is a straight-line affair. By contrast, evolution, in the manner Darwin proposed, requires vast swaths of time—geological time. The gradualism that Lyell advocated—"the slow agency of existing causes"—works equally well as a driver of geological or biological change. This was the foundation on which Darwin built what the philosopher Daniel Dennett called "the single best idea anybody ever had."[48] Without Lyell, Darwin the man could not have become Darwin the historical figure that we know.

# 7 Making Sense of Monsters

On October 2, 1836, the HMS *Beagle* came to anchor in Falmouth, England, nearly five years after its departure from Devonport. The world to which Charles Darwin returned had changed. Lyell's *Principles of Geology* had convinced not only Darwin but also a great many other scientists. Buckland dropped his reliance on the Great Deluge, and even some strident catastrophists, such as Darwin's old mentor Jameson, began to soften their objections to the gradualistic approach of Hutton and Lyell.[49]

If the Earth had an ancient past, well then, what happened in it? As an awareness of our planet's antiquity grew, the fields of geology and paleontology gained more attention. The spectacular discoveries of Buckland and Mantell, and of Cuvier and others in Europe, were capturing the public's imagination. Mantell opened his own museum in Brighton, fifty miles south of London, which became one of Europe's most famous collections.[50] On the coast of Lyme Regis, on the English Channel, Mary Anning, famous for her discovery of fossilized remains of the marine reptiles *Ichthyosaurus* and *Plesiosaurus*, ran a popular rock and fossil shop, Anning's Fossil Depot. Tourists were particularly fond of purchasing the ancient ammonite shells, which she incessantly beat from the clutches of Jurassic limestone cliffs. The tongue twister "She sells seashells on the seashore" is about her.[51]

With interest mounting, in 1837 the British Association for the Advancement of Science commissioned anatomist Richard Owen to compile a *Report on the Present State of Knowledge of the Fossil Reptiles of Great Britain*. Near the conclusion of his work, in August 1841, Owen traveled to Plymouth to present his results to the association. He did not mention dinosaurs, though; he hadn't "invented" them yet. But sometime during the fall of 1841, he made a striking realization. *Megalosaurus*, *Iguanodon*, and *Hylaeosaurus*, apart from their great size, all held an unusual characteristic in common: the vertebrae inside their hips, known as their sacral vertebrae, were fused to one another. An adaptation for terrestrial locomotion, he thought.[52] This, Owen reasoned, was cause for grouping them together.[53] And if they were grouped together, they would need a name.

He revised his report and published his results in early April 1842, the *Report on British Fossil Reptiles*.[54] Buried deep within volume 2, on page 103, was this line describing the saurians of Buckland and Mantell: "The combination of characters . . . will, it is presumed, be deemed sufficient ground for establishing a distinct tribe or sub-order of Saurian reptiles, for which I would propose the name of Dinosauria [meaning fearfully great lizard]. Of this tribe, the principle and best established genera are the *Megalosaurus*, the *Hylaeosaurus*, and the *Iguanodon*; the gigantic Crocodile-lizards of the dry land."

Dinosaurs were now a *thing*. Owen recognized in these three giant beasts characteristics that set them apart from all other reptiles. But "Crocodile-lizards"? *Megalosaurus*, *Iguanodon*, and *Hylaeosaurus* were amazing fossils in that they opened a window into the study of one of the most awe-inspiring group of creatures to have ever lived. As anatomical specimens, however, they formed a poor collection. Owen was a superb anatomist, but he had such scant remains, he was not able to discern the inherent vigor and power of dinosaurs, which sets them apart so dramatically from their crocodilian cousins.[55]

Burying the lede, as he did, one wonders if Owen had any inkling of the enduring fascination and enthusiasm with which the public would embrace dinosaurs. He would soon find out.

*Megalosaurus*, *Iguanodon*, and *Hylaeosaurus* became household names. By 1852, the idea of dinosaurs had so permeated the Victorian zeitgeist that Charles Dickens himself, writing in *Bleak House*, clearly assumes his readers will understand his reference to *Megalosaurus*: "Implacable November weather. As much mud in the streets as if the waters had but newly retired from the face of the earth, and it would not be wonderful to meet a Megalosaurus, forty feet long or so, waddling like an elephantine lizard up Holborn Hill."

In 1851 crowds packed London's Hyde Park for the Great Exhibition of the Works of Industry of All the Nations, essentially the first World's Fair. Its centerpiece was an architectural triumph, the 990,000 square foot Crystal Palace, which showcased the latest technological achievements in steel and glass manufacturing. When the palace was moved from its temporary exhibit grounds in Hyde Park to a permanent home in South London in 1854, a dinosaur garden containing life-size sculptures was added to the design—the first Jurassic Park.[56] The artist Benjamin Waterhouse Hawkins was commissioned to re-create the three known dinosaur species. Owen would advise on the anatomy, and the results would fully reflect his views. The dinosaurs were portrayed as brutish, plodding reptilian creatures, the "Crocodile-lizards" of his mind's eye. Remarkably, the sculptures still exist today and have become a point of pilgrimage for dinosaur enthusiasts around the world. I visited them while writing this book, and found them charmingly anachronistic. The *Megalosaurus* resembles the love child of a Gila monster and a buffalo. The pigeons perched upon its head appeared wholly nonplussed by their giant Jurassic cousin. The *Hylaeosaurus* could be described as a prickly, walleyed lizard. And the *Iguanodon* looks like a genetic mishap involving a corpulent crocodile and a rhinoceros. (Mantell had mistook the spiky thumb of *Iguanodon* for a horn, an error that Owen carried over into its reconstruction.)

The Crystal Palace dinosaurs proved to be wildly popular. Through the end of the nineteenth century, more than a million people a year came to gawk at the gargoyle-like effigies returned from an ancient land. There had never been anything like them, and they were the most viewed scientific exhibit in the world.[57]

Hawkins' sculptures were fully fabricated, 3-D reconstructions, depicting the outward appearance of dinosaurs. Decades would pass before the Royal Scottish Museum in Edinburgh exhibited Europe's first mounted dinosaur skeleton, showing the actual anatomy of the bones. The specimen displayed would not be cast from Buckland's or Mantell's collections, though, or even from fossils from Europe; these were far too fragmentary to create a reasonable skeletal mount. No, the first dinosaur skeleton displayed in Europe would come from New Jersey.

In 1858 William Parker Foulke, a Philadelphia attorney, was summering in the pretty little farming village of Haddonfield, New Jersey. There he had heard stories of large bones exhumed from the ground by diggers of marl, an ancient kind of fertilizer. He snooped about and found the site. It was a farm belonging to John Hopkins, who told Foulke that souvenir seekers had carried off the bones twenty or so years before. Hopkins, not having much interest in such things, told Foulke he was welcome to poke around. Foulke assembled a team, rediscovered the overgrown marl pit, and began digging. After excavating down about ten feet, they hit a layer of fossil shells and bones—large bones.[58] The stories were true.

Foulke alerted Joseph Leidy, a polymath scientist at the nearby Academy of Natural Sciences of Philadelphia. Leidy and his men joined the dig, and the combined team labored into the fall. In total, they recovered forty-nine bones and teeth from a single creature. They knew they had uncovered the most complete dinosaur skeleton the world had yet seen, and they must have been excited.

Back in Philadelphia, Leidy set right to work describing the bones. It appeared to him to be "a huge herbivorous saurian," related, he knew, to Mantell's *Iguanodon.*[59] But unlike the fragmentary British dinosaurs, whose anatomies had been obscured for want of better remains, the Haddonfield dinosaur was relatively complete. The fog of uncertainty that had shrouded these beasts since their discovery began to clear. To be sure, Leidy's view of dinosaurs was still a murky one, but this specimen began to clarify the anatomical pea soup through which Owen had tried to peer. Most striking to Leidy were the proportions of the limb bones. The upper arm bone (the humerus) was only about half the length of the thighbone (the femur). This animal could not have been a clunky quadruped, as Owen had imagined *Iguanodon*. This plant eater was upright—a biped. Leidy likened it to a kangaroo, standing erect and forming a tripod between its hind feet and tail. The animal that he envisioned does not quite comport to our current understanding of these creatures. But unlike Owen's "Crocodile-lizards," you would recognize Leidy's plant eater as a dinosaur, for it had a hint of the vigor and power.

Leidy named the new species *Hadrosaurus foulkii*, meaning literally "Foulke's bulky lizard," though I have to believe that the name was contrived as a double entendre, giving a nod also to Haddonfield, New Jersey. By the end of 1858, Leidy had presented his new dinosaur to the members of the Academy of Natural Sciences, establishing the New World's first named dinosaur.

That same year, Benjamin Waterhouse Hawkins, by then famous for his Crystal Palace creations, immigrated to Philadelphia, the new epicenter of dinosaur paleontology. Hawkins marveled at Leidy's collection of *Hadrosaurus* bones and asked for and received permission to cast each one. In European museums, dinosaur bones had been simply laid out in cases for the public to view. For those not trained in anatomy, this paints a poor picture of the creatures that once existed. Hawkins dreamed of creating a skeletal mount of *Hadrosaurus*, one that would reveal to the laity its towering size and breathtaking grandeur.

In his workshop, he set about sculpting each of the missing pieces of *Hadrosaurus*, including the entire skull, which was purely speculative. He created an iron armature to which the bone casts would be fitted. Guided by Leidy, Hawkins' *Hadrosaurus* was the antithesis of his Crystal Palace dinosaurs; it was bipedal, upright, and appeared to be poised for action. This was no paunchy, half-comatose Crocodile-lizard. This was an alert, vigorous creature, a beast of consequence in its landscape.

The buzz over the discovery of *Hadrosaurus*, and interest from major museums, caused the marl miners in the vicinity to become more attentive to the occurrence of fossil bones. Over the remainder of the nineteenth century, dinosaur and marine reptile fossils began popping up in the many marl pits that dotted the bucolic Cretaceous countryside of southern New Jersey.

In 1866, miners in a marl pit near the sleepy crossroad village of Barnsboro, New Jersey, uncovered the bones of a large and savage predator. The quarry superintendent, J. C. Voorhees, allowed Edward Drinker Cope, Leidy's young protégé, to take the bones back to the Academy of Natural Sciences. Upon examining the remains, Cope immediately recognized the specimen as a meat-eating dinosaur related to Owen's *Megalosaurus*. Voorhees' diggers had recovered more than two dozen bones of a single individual and an associated collection of teeth. It was the first skeleton of a meat-eating dinosaur found and described in North America. Cope noted that its arms were relatively short compared with its legs. Another bipedal dinosaur, he thought, like *Hadrosaurus*, only this time, a carnivore. It was large. Its hips would have stood as high as a draft horse. It had powerful legs and a powerful tail. Its teeth were flattened and serrated; a mouth full of steak knives. Its arms ended in meat hooks, eight-inch sickle-claws of "remarkable and destructive use,"[60] said Cope, three per hand, at the ends of its outspread fingers. A one-and-a-half-ton food processor on legs. An animated abattoir, stalking the ancient coastline of New Jersey.

*Tyrannosaurus rex* is, without contest, the most famous dinosaur. It is, however, not the only tyrannosaur. More than two dozen species are known, grouped into the superfamily (in Linnaean terms) Tyrannosauroidea, and J. C. Voorhees's men had just found the very first one. Cope named the beast *Laelaps*, but the name, it turns out, was preoccupied by a tiny species of mite, and was changed eleven years later, to *Dryptosaurus*. Cope could not know, of course, that he was naming the first described tyrannosaur. The discovery of *T. rex* was nearly four decades off, and the grouping of related predators into a superfamily would come a century after his description. It is a delicious morsel of paleontological history, though, that the first tyrannosaur described was discovered not in a rugged Wyoming basin, or on the windswept plains of Montana, or in the badlands of South Dakota, but among the peach orchards and tomato farms of southern New Jersey.

Two years after the *Dryptosaurus* discovery, Hawkins was putting the final touches on his *Hadrosaurus* mount. It was placed on display in the Academy of Natural Sciences in 1868. Once again, Hawkins' artistry thrilled the public. The exhibit was a smash hit. Attendance doubled. Dinomania had hit the United States. Overwhelmed, the academy started charging admission to quell the thongs. It didn't work. Ultimately, it would have to relocate to a new, larger building (the academy's current home), to accommodate the public's insatiable desire to stand in a room with a dinosaur. Buoyed by popular demand, Hawkins would cast duplicate copies of his sensational *Hadrosaurus* for Princeton University, the Smithsonian Institution, and the Royal Scottish Museum.

Less than a year after *Hadrosaurus* was announced, Charles Darwin published *On the Origin of Species*, a book of unmatched scientific importance and a work that, to this day, underpins all biological understanding. In it he painstakingly laid out his astoundingly thorough argument for transmutation (evolution) by natural selection. Key to his rationale was a recognition of the variability that exists naturally in all populations of organisms. Darwin didn't know about genes when he published, although at the time Gregor Mendel, the Austrian scientist and Augustinian friar, had experiments underway that would unravel the mystery. (Monks tend to keep to themselves.) Darwin reasoned, though, that there must be some mechanism of inheritance, and through this engine of transmission, traits would be passed on to succeeding generations. Those able to pass along the most copies of their traits were, he thought, the fittest. Using a concept borrowed from the economist Thomas Malthus, Darwin deduced that in an environment in which resources are constrained, variants within a population will achieve differential success. In other words, natural selection acts as a screen, filtering out the characteristics of less fit individuals and allowing more fit individuals to propagate a greater proportion of their traits into the future.

With natural selection ever acting as the dispassionate arbiter of success, driving a population to change in one way or another, Darwin hypothesized that if a group of individuals was reproductively isolated—stranded on an island, for example—they would eventually carve out their own evolutionary pathway, separate from that of their disconnected kin. Given enough time, they would become their own species, a process, he saw, that would produce a branching pattern, a tree of life. View a tree in retrospection, though, and what do you see? A coming together of branches, a reduction of species right down to a single trunk—common descent, as Darwin put it.

Now the pieces were in place to truly understand the countless past worlds that have been our planet. Hutton and Lyell gave us an ancient Earth, one old enough for "the slow agency of existing causes" to produce the wondrous physical landscape we see around us. Darwin took this uniformitarian view and applied it to living things. Life, he saw, was a grand struggle, a vicious competition for resources that favored only the fittest. Paleontological discoveries would now be viewed through this lens, with fossil species representing the victors of their age in the competition to occupy the future.

By the turn of the twentieth century, the Darwinian view was fully infused into our conception of ancient life. An 1897 painting of *Dryptosaurus* dramatically illustrates the progress that had been achieved over the preceding half century.

On his deathbed, Edward Drinker Cope received a visit from a young Brooklyn artist, Charles R. Knight, who had

For most of civilized history, we humans had considered ourselves the divinely anointed masters of our planet, beings created apart from the plants and animals. With a birthright of dominion over all that we saw, we had placed ourselves at the center of Creation. But as we began scrutinizing rock formations and unearthing dinosaur bones and fossils of every kind, the subjects of our curiosity transformed from mystery into mirror.

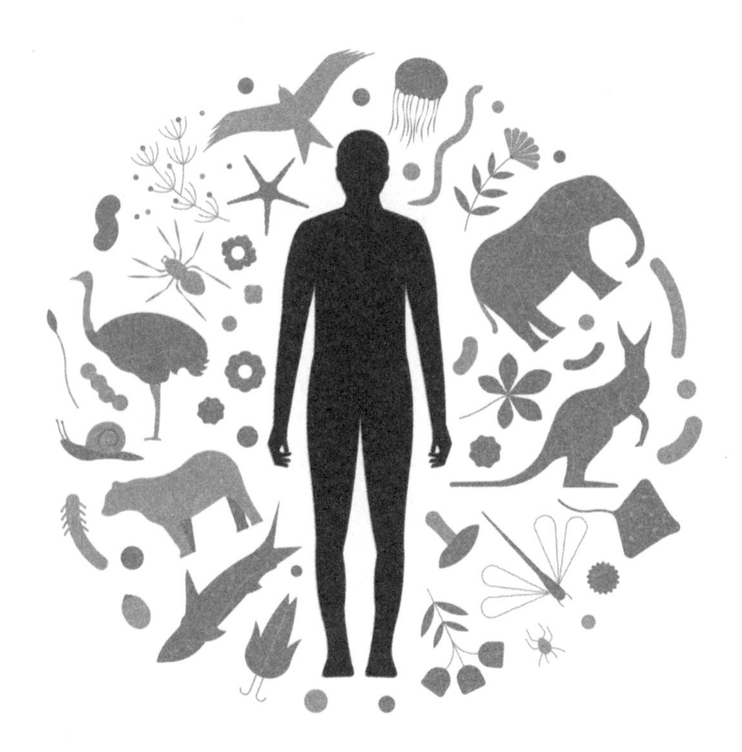

With our discovery of deep time, evolution, and common descent, we saw that far from being the reason why our planet exists, we are but one among multitudes, each one being enormously lucky, each one contingent on so much else, and each one here but for a fleeting moment in the vast tapestry of deep time.

obtained a commission to paint a series of prehistoric creatures. Knight wanted to reconstruct *Dryptosaurus*, and Cope, in what would be his last contribution to science, agreed to advise. The collaboration produced a stunning result: the world's first view of dinosaurs as vigorous, nimble, and highly competent creatures. The painting, entitled *Leaping Laelaps* (for *Laelaps* was Cope's original name for *Dryptosaurus*), features a pair of *Dryptosaurus* engaged in mortal combat. In it the defender lies recumbent, with a terrible phalanx of dagger-like claws directed at the furious aggressor. The attacker is uncompromising, sailing through the air, coiled like a spring, ready to release a deadly explosion of evisceration on its foe. Along with Knight and Cope, Darwin could well be considered the third collaborator in this Mesozoic masterpiece.

*Leaping Laelaps* is a prescient portrayal of dinosaurs, the likes of which would not be seen again for another hundred years. The painting is imbued with the Darwinian view of life: a competitive and savage struggle for survival. In Knight's scene, there will clearly be a winner and a loser, or possibly two losers. (I think we can rule out a win-win situation.) Either way, the ability of these two individuals to transmit their genes into the future lies at stake. Why are they battling? Knight leaves that up to us to contemplate. Combat over a mate, perhaps? They might be laying mortal claims upon a territory. They might be embroiled in a deadly argument of ownership over a particularly succulent duckbill carcass lying just beyond on the forest floor. Whatever the cause, we see in this struggle the incremental functioning of the existing causes, mindlessly driving evolution toward increasing fitness.

Thinking back on the state of human understanding just a few hundred years prior, the Hutton-Lyell-Darwin revolution overhauled not just our view of the world but also our view of ourselves. Was the Earth not created for us and sovereignty over it given to us? Was our place upon this planet not inevitable, not foreordained? Are we truly no more than a late-sprouting twig on the vast and bountiful tree of life?

A seventeenth-century farmer turning up a fossil shark tooth might have pondered the mysterious, mystical forces that spontaneously generated such a sublimely shaped object, or he might have silently recited a little prayer in his head, giving thanks for the miraculous handiwork of an all-powerful creator. But he would not have seen the biological thread tying it to him. He would not have meditated on the antiquity of the object and experienced the humility that the contemplation of deep time so often brings to those who consider it. It's stunning to think that our species, related to every other living thing on the planet, and the product of 3.8 billion years of evolution, on a 4.5-billion-year-old world, had convinced itself of just the opposite. We had told ourselves that this planet was our temporary domicile and that we were created, quite recently, as something apart from it—special beings, endowed with a right of dominion over all that we saw.

By the end of the nineteenth century, though, paleontology, geology, and evolutionary biology had forged a new way for humanity to understand itself and to contextualize the world around us. By removing us from center stage in the drama that is Earth history, we began to see ourselves as part of nature, as part of this world. For some, this view is unsettling, ego bruising, and out of line with deeply cherished beliefs. For many others, though, the discovery of deep time, evolution through natural selection, and common descent brings with it a sense of connectedness to all other living things and a humbling feeling of smallness of being. Within the incomprehensible ocean of time, now revealed to us, our entire species amounts to no more than a single roiling white cap that bursts forth upon a wave, effervescing for a brief, sparkling moment with the fleeting lives of billions.

# 8 The King

Dinosaurs were tiny and huge. They were skittish and ferocious. Fast and slow. Runners, walkers, climbers, flyers,[61] and sometimes swimmers.[62] They were solitary and gregarious. Nocturnal and diurnal. Meat eaters and plant eaters. Hunters, scavengers, grazers, and browsers. Drab, colorful, scaled, and feathered. Endemic and widespread. Territorial homebodies and peripatetic wanderers. Dinosaurs were tropical, temperate, and polar. They lived on beaches and lake shores, alongside rivers, and in mangroves, fern lands, forests, bogs, floodplains, and on mountain sides. They were astoundingly adaptable.

The staggering array of dinosaur shapes, sizes, abilities, and behaviors exemplifies the power of evolution. Yet they were improbable. Worlds don't have to have dinosaurs any more than they have to have humans. We're all freaks of nature, us species, in that everything has to line up just right for us to be. It's remarkable that the same mechanisms of natural selection, amplified over deep time but modified by circumstance, led to both dinosaurs and us. Their story is not our story, though we share the first few chapters, yet the common themes of evolution—such as efficiency, resilience, adaptation, competitiveness, and dispersal—carry through and resonate with our own experience.

Despite their dizzying variety, their wide range of habitats, and their manifold behaviors, the cornucopian richness of these animals and their incredible span across time is often collapsed into the phrase "the dinosaurs." How did the dinosaurs go extinct? Were the dinosaurs warm- or cold-blooded? How did the dinosaurs breed? Did the dinosaurs take care of their young? Did the dinosaurs have feathers? It is impossible to answer these questions succinctly because, except on the broadest scale, there is no "the dinosaurs." Today there are over eighteen thousand bird species, and thus far paleontologists have described over a thousand non-avian dinosaur species. There must have been many, many more. Some lie waiting to be discovered, and many others, perhaps most, have been lost to the twin ravagers of Earth history: geology and time. From our vantage point in the present, the true dimensions of dinosaur biodiversity will never be revealed. Even so, the fraction of dinosaurs that have been levered out from beneath the shroud of deep time comprise a breathtaking menagerie of remarkable creatures. None of which is more spell-binding than the charismatic king of dinosaur pop culture, *Tyrannosaurus rex*.

A decade ago, I attended the opening of a big-budget arena show, inspired by the acclaimed BBC documentary *Walking with Dinosaurs*.[63] The packed Spectrum sports arena, on South Broad Street in Philadelphia, was rocking with as many as fifteen thousand crazed fans. It could well have been a Bruce Springsteen concert, except that half the revelers were under four feet tall and pounding cotton candy, not beer. The $20 million stage show featured computer animations, physical animatronics, and puppetry, which combined to create an impressively realistic and, at times, thrilling experience.

Backstage after the show, I saw that the dinosaurs possessed intricate and subtle details. Individual scales, each hand-painted, giggling flaps of skin, serrated teeth on some, and eyes that seemed to look back at you all combined to give the effect of being in a room with actual dinosaurs. There was *Brachiosaurus*, *Stegosaurus*, *Torosaurus*, *Plateosaurus* and her hatchlings, *Utahraptor*, *Ankylosaurus* (bigger than I had imagined from its skeleton alone), and *Allosaurus*, a childhood favorite. Then it was time to meet the king. Though I might have been a bit starstruck, I was certain that I was not seeing double. Yet there was not one but two *Tyrannosaurus rex* looming over me, though I was sure I remembered only one from the performance. I grabbed the head of puppetry and asked him why there was a second *T. rex*. The additional tyrannosaur cost a million dollars to make, he told me, "but we need it. It's the understudy." If one of the other dinosaur cast members breaks, he explained, it wouldn't stop the show. But showing up without a *Tyrannosaurus rex*? No way. "That would cause a riot," he said. "We might as well just go home." Such is the popularity of the world's most beloved dinosaur.

The holotype of *T. rex,* meaning the defining specimen, was discovered in 1902, by the renowned American Museum of Natural History collector Barnum Brown, who unearthed its skeleton in the sixty-six-million-year-old Hell Creek Formation of Montana. The remains were described in print three years later by paleontologist Henry Fairfield Osborn, who coined its iconic name, which means "tyrant lizard king"—one of the great species names of all time.

Many of the best known dinosaurs are also among the first discovered. *Allosaurus* and *Stegosaurus* (both published in 1877), *Brontosaurus* (1879), *Triceratops* (1889), *Tyrannosaurus rex* (1905), and *Ankylosaurus* (1908) have all been banging around in pop culture longer than almost all other dinosaurs. This, I think, largely accounts for their popularity and name recognition.

*Tyrannosaurus*, though, is deserving of its unsurpassed measure of fame for many reasons. Not least of which is that it's huge! A fleshed-out *Tyrannosaurus* would have weighed eight or nine tons and would have stretched forty-five feet from its snout to the tip of its tail. With its large eyes and excellent binocular vision, it could have reckoned the distance of objects with the skill of a baseball outfielder. One study suggests that *T. rex* possessed thirteen times the visual acuity of humans, greatly surpassing even that of modern-day avian raptors.[64] To be "tyrannosaur-eyed" is superior to being merely "eagle-eyed." The skull of *T. rex* shows that it housed well-developed olfactory senses. In the film *Jurassic Park,* standing motionless before a *T. rex* is sufficient to avoid its notice. In the actual Cretaceous Period, when *T. rex* lived, this tactic would have likely led to your ingestion. If you stood statuesque in the vicinity of a *T. rex,* it would smell you like a hound dog and devour you like a monster, for its mouth was packed with as many as sixty teeth, nine-inch-long flesh-penetrating daggers. "Killer bananas," as the paleontologist Robert Bakker called them. This deadly collection of weaponry was driven into its prey by outsized jaw muscles that powered the most forceful bite ever possessed by a land animal. *T. rex* could shred its victims and crush their bones, slurping down in a single gulp enough meat to stock a butcher shop. It preyed upon whatever it could catch and probably scavenged whatever else it could find. Its taste for all things meat seems to have known no bounds. Evidence of cannibalism is surprisingly common for *Tyrannosaurus*. It makes one wonder what cautious negotiations must have proceeded their nuptial parings.

Few, if any, animals possessed armor tough enough to withstand the bite of *Tyrannosaurus*. Hunkering down in the face of a *T. rex* onslaught would be a perilous proposition. But running for your life would be no less hazardous. *T. rex* was fast. Most studies on *Tyrannosaurus* locomotion suggest that it could outpace even the speediest Olympic sprinter. So, how might you survive if you ended up on the Isla Nublar with a nine-ton *T. rex* bearing down on you? Let's review. Standing motionless equals death. You'd be smelled. Fending off *T. rex*?

Death. Attempting to outrun *T. rex*? Death. Jumping in the water? Death. *T. rex* could swim.[65] Taking flight? Death. (You can't fly.)

But wait, there is one strategy that might keep you alive. Humans are relatively agile and can pivot their bodies in a fraction of a second. *Tyrannosaurus rex*, though, would have been a laborious turner. Across its forty-five-foot-long body, it held a great deal of mass far from its center of gravity. Imagine hefting a telephone pole across your shoulders and trying to pivot. Conservation of angular momentum would make this operation ponderously slow. So, what to do if a *T. rex* is about to snatch you from behind and gobble you up like a meat cookie? Make a U-turn! Doubling back on a marauding *Tyrannosaurus* is a terrifying prospect. But as poor as this option is, attempting to outmaneuver this beast is your best bet to stay alive. Admittedly, though, you would find me at the back of the line to test this particular hypothesis.

The evolution of *T. rex* into a massive megapredator is a testament to its hunting proficiency. Its mere existence in the rock record gives silent acknowledgment to all the morning-chilled lizards, distracted mammals, stumbling dinosaurs, and turned-upon members of its own species that never got away. The sum of its qualities made *Tyrannosaurus* an extremely dangerous creature to any land animal substantial enough to warrant its notice and unfortunate enough to share its domain.

But what about those arms? Those dinky, withered, Lilliputian *T. rex* arms. Arms no longer than your arms, on a creature the size of a bus. They must have been a terrible drawback. The Achilles' heel of arms. Picayune appendages, unworthy of a mighty ruler. Second-rate limbs maligned by common mammals in countless Internet cartoon memes. "*T. rex* hates push-ups." "If you're happy and you know it . . . aw, never mind." "*T. rex*, worst DJ ever." "I love you this much," says one *T. rex* to another, with open arms. "That's not very much," replies its downcast partner. "Can you pass the salt?" one *T. rex* asks another. "Don't be mean," replies the other. A sad *T. rex*, dreaming of big, powerful arms but unable to reach the genie's lantern on the ground below. A *T. rex* stretching its reach with a set of late-night-TV extendable grabbers, declaring itself to be "unstoppable!" A tyrant lizard king sitting on a throne, of sorts, indisposed and forlorn, unable to wipe its own bum. Oh, the indignities!

But what if I told you that the puny arms of *T. rex* represent one of its greatest strengths? What if I told you that its diminutive limbs were, in reality, a key adaptation that allowed it to terrorize its landscape and dominate its ecosystem? Could this be true?

Yes, it could. It's counterintuitive, but natural selection may reward seemingly deleterious adaptations if those traits enhance the fitness of the individuals carrying them. For example, in recent human evolution, we have seen the power of our jaws diminish, our teeth shrink, our appendix atrophy, and our protective covering of fur lost. In terms of our ability to survive in the wild, these may not seem like positive changes. But it's important to realize that fitness is agnostic on methods and reflects only reproductive success, however achieved. To understand how losing certain abilities may have enhanced our own fitness, we can look to the much maligned forelimbs of *T. rex*.

Whether or not the arms of *Tyrannosaurus* look out of proportion to us is wholly irrelevant. As the astrophysicist Neil deGrasse Tyson is fond of saying, "The universe is under no obligation to make sense to you." Natural selection has no aesthetic and is completely devoid of any sense of humor. It is nothing more or less than an insensate arbiter of efficiency, as measured by fitness. The qualities that lead to enhanced fitness are sometimes obvious (being a fast rabbit) and sometimes not (being a legless lizard).

To illustrate this point, we can look at the evolution of vision. Touch and taste are the most parochial of the senses, useful, but dangerous to employ, requiring bodily contact. The other senses—vision, hearing, and smell—are like superpowers: an array of remote sensing devices collecting information about threats and opportunities from afar. Sight, in particular, has proved to be so beneficial to organisms that it has evolved independently fifty to a hundred times.

Seeing is such an advantage that even a poor eye can be useful and can tip the balance in favor of an individual. The blue sea star (*Linckia laevigata*), for example, has eyespots at the tip of each of its five appendages. The primitive eyes of this starfish have no lenses and can form only a crude image, the equivalent of about fourteen pixels across.[66] Yet these rudimentary eyes allow the blue sea star to discern light from dark, a critical piece of information when assessing which way is up, determining the time of day, or when looking for just the right hiding place. Importantly, researchers recently discovered that the blurry image assembled by the blue sea star's eyespots allows it to visually navigate back to its reef habitat after being dislodged from it—a life-saving ability. For the blue sea star, living in a dim and out-of-focus hazy world is safer than living in total darkness, and the energy it puts into eyespots reaps substantial rewards.

But sometimes evolution blinds a species; takes away its eyes. How can this be? If vision is so beneficial, why would natural selection drive a species to lose its eyes?

We see this occur in some troglobitic species: those adapted to living underground. For animals dwelling in total darkness, eyes have a cost but no benefit. The fuzzy image that guides the blue sea star makes eyes worth having for the tiny advantage it confers. But take away the light, and the eyes have to go. It's use it or lose it. In evolution, small changes in the probability of success stack up over time and have big effects. It's the same principle that keeps casinos in business. It may seem like you have an even chance of winning, but you don't. Each game has a built-in house advantage that doesn't matter much over a few bets but keeps the casino winning (and you losing) over time. Setting the odds just slightly in the casino's favor or in a phenotype's[67] favor is the key to success.

In a recent study, a team of researchers examined the energy costs associated with vision for seeing cavefish versus blind cavefish. To do this, they recorded oxygen consumption for cavefish eyes and the vision-related portions of their brains.[68] The blind cavefish consumed 15 percent less energy than their sighted cousins did. It turns out that nerve cells (neurons) and photoreceptive cells are particularly energy hungry. Additionally, blind Mexican cavefish (*Astyanax mexicanus*), having no need to process visual information, get by with significantly smaller brains, another energy savings. While it might be nice to have eyes for special circumstances, blind cavefish do better by losing them and conserving the energy instead. With natural selection, "Use it or lose it" is the rule by which the cost, risk, and benefit associated with every body part is judged.

By now, you've probably guessed that this rule pertains to the tiny arms of *T. rex*. Its ancestors had arms that were bigger and more powerful, but they came at a price. It costs energy to make big arms and it costs energy to maintain them. There is also risk associated with having arms. Arms have bones that can break and tissue that can go pathological. Infections cannot take hold in an arm that isn't there. Cancer cannot spread through a limb that never was. No calories are consumed by an arm that failed to be. Arms are not free. So why have them? Well, for some animals, they are essential. It's hard to imagine any species of primate, ourselves included, being successful at the things primates do without a well-functioning set of arms. Clearly, for primates, the benefits of having arms outweigh the costs of growing and maintaining them, and the risks associated with having them. I know these risks firsthand.

I am typing these sentences with a left appendage girded by a stainless steel bar nine inches long, fastened down with nine large screws that were twisted into my ulna (the larger forearm bone) by a surgeon—a trophy from my mountain biking days. If I lived in the nineteenth century or before and suffered this same injury, I would probably have an atrophied, fairly useless left arm, or I quite possibly might have died from shock or infection at the time of the injury. From a biological standpoint, bearing arms is not without risk.

Now imagine a population of *T. rex* ancestors. They were long-armed, like *Dryptosaurus*. Of course, within this group, there were physical variations; different phenotypes. Among the differences, some individuals possessed forelimbs a bit longer or shorter than most. Over time, evolution favored the shorter-limbed varieties—and knocked off a digit as well. The result was *T. rex*, resplendent with its imposing stature, menacing teeth, and burly thighs, but with puny arms tipped with two witchy fingers. A bit of comic relief in what is otherwise the stuff of nightmares.

Tyrannosaurus rex, the most iconic of all dinosaurs, possessed the most powerful bite of any land animal in history. Counterintuitively, the key to its power hinges on those tiny, much ridiculed arms. Arm muscles, it turns out, compete with neck muscles for attachment space in the shoulder. A powerful bite requires a big head, which requires big neck muscles for support. It's a zero-sum situation—too much arm, not enough bite.

But why were the longer arms of *Tyrannosaurus rex* ancestors a competitive disadvantage? Michael Habib, a paleontologist at the University of Southern California, points out that even with relatively longer arms, predators built like those that gave rise to *T. rex* could not reach their mouths with their hands. They "probably couldn't even see their arms while on the hunt," he adds.[69] With little utility, the "Use it or lose it" rule kicked in, and tyrannosaur arms withered away. In an unexpected way, this may have opened the door for *T. rex* to become the most terrifying of all dinosaurs.

*Tyrannosaurus rex* possessed the most powerful set of jaws ever attached to a land animal. Robust jaws necessitate huge jaw muscles attached to a large head. And an immense head requires powerful neck muscles for support. Habib notes that neck muscles and arm muscles compete with one another for muscle attachment space across the bones of the chest and shoulders.[70] Without sufficiently large neck muscles to support a sufficiently large head, *T. rex* would be denied its most lethal weapon: its devastating bite. Because huge forelimbs were not necessary for its success, a trade-off was possible: a reduced set of arm muscles for a bulging set of neck muscles. As the arms of *T. rex* diminished, its bite became more and more devastating. It's true, *T. rex* can't do push-ups, answer the phone, or put on its hat. But it can crack open the skull of a two-ton duckbill dinosaur with a single bone-shattering chomp. Perhaps now you'll see those arms as a whole lot less comical in view of the murderous mouth they make possible.

The tale of *T. rex* tells us that small things matter, even if you're huge. Just as casinos rely on setting the odds ever so slightly in their favor, the miniscule benefits accrued by tyrannosaurs that had slightly smaller arms helped give rise to a predator of unparalleled destructive force. It's a reminder that strengths and weaknesses may come in pairs—"twins of the same womb," as the poet Marge Piercy puts it.[71] Think of the physical weakness of passive resistance and how it can build a moral force that can stop tanks and topple governments. There is a reason why despots around the world fear feeble acts of resistance. Given the right moment and a little amplification, the hunger strike, the sit-in, and the silent protest can spawn the twin of weakness: a force that can bite back with the ferocity of a *Tyrannosaurus rex*.

# 9 Champions

As amazing as *T. rex* was, it was far from unique in its uniqueness. Every dinosaur species is a tale of success, of beating the odds, of emerging victorious, where others had fallen. Every dinosaur species possessed a never-before-seen set of attributes that allowed it to carve out a niche from its landscape. Every dinosaur species, by its own agency, descended from a lineage that survived from the dawn of life, through nearly four billion years of evolution, despite unrelenting, withering competition from other species and the vicissitudes of a capricious planet. The 165-million-year reign of the non-avian dinosaurs was a riot of adaptation played out on every continent and in every terrestrial environment. Once our eyes were opened to deep time and evolution, the results would surprise us over and over again.

There was *Spinosaurus aegyptiacus*, a dinosaur as long as *T. rex*, with a crocodile-like skull, long gaff-hook arms, and a six-foot sail running down the length of its back. It was a piscivore—a fish eater—and hunted in the Gordian knot of tidal channels that twisted and sprawled through the ancient mangrove coast of North Africa. A 2014 study postulated that it was semiaquatic, with flat feet that may have been webbed to help propel it through the water.[72] Many tiny nerve openings along the front of its skull show that it had a probe-like, sensitive snout. Its nostrils were retracted halfway up toward its eyes. Together these features made it perfectly adapted for mucking about for a meal in the muddy tidal creeks that it terrorized. With its limbs roughly even in length, an adaptation for paddling, the researchers speculated that, on land, *Spinosaurus* walked on all fours, unlike all other meat-eating dinosaurs. Paleontologist Paul Sereno of the University of Chicago quipped, "It's like a cross between an aquatic bird and a crocodile."[73]

Bizarre in other ways was *Anzu wyliei,* a species of oviraptorosaur cobbled together from three specimens found in the badlands of North and South Dakota.[74] *Anzu* looked like a cross between a five-hundred-pound chicken and a drawer full of cutlery. It stood as tall as a person and was lavishly feathered. It menaced the ancient Dakota bayous with its sharp claws and toothless beak that snapped together like a pair of meat cleavers. "It had a fiendish appearance; it wouldn't be hard to imagine *Anzu* terrifying children in a Jurassic Park kitchen," says the lead author of its description, Matthew Lamanna of Pittsburgh's Carnegie Museum of Natural History, who dubbed *Anzu* the "chicken from Hell."

The oldest known bird is *Archaeopteryx*, from Jurassic lake deposits in Germany and Spain. The feathers that adorned the much younger *Anzu*, however, were not passed down from *Archaeopteryx* or any other bird. They were inherited from non-avian feathered dinosaurs that pre-dated the birds of the Jurassic. The first function of feathers was not flight; it appears to be insulation. Because energy that is lost to the outside world as heat is not available to power growth, insulation became more important as dinosaur metabolisms rose.[75] For some dinosaurs, downy feathers provided the necessary thermal barrier. But among the non-avian dinosaurs such as *Anzu*, a second type of feather evolved: the pennaceous feather. In birds, these feathers are needed for flight. But what was their function in a pedestrian like *Anzu*?

A clue comes from studying the vision of the closest living relatives of non-avian dinosaurs: birds and crocodiles. Their eyes contain four sets of color receptors, ones for red, green, blue, and ultraviolet light. This is the ancestral state for land vertebrates. By contrast, human eyes possess only the first three. There is no bluer than blue for us. We lost this ability during the long nights of the Mesozoic, when our shrew-like ancestors adopted nocturnal ways to avoid the sharp gaze of hungry daytime dinosaurs. In their dimly lit world, the "Use it or lose it" rule trimmed the excess receptor, and we lost our ability to see ultraviolet light.

With their superior color vision, it is no coincidence that some birds produce spectacularly vivid plumage. This is where pennaceous feathers play a second role. Downy feathers are great for insulation, but because they do not form a coherent light-reflecting surface, they cannot produce the shimmer of iridescence that dances across the bodies of purple grackles, illuminates the shockingly blue necks of peacocks, and transfigures flitting hummingbirds into sparkling garden ornaments. These special effects of nature are produced by the smooth, interlocking surfaces of pennaceous feathers, which are thought to pre-date the evolution of flight. Their main function, when they appeared in non-avian dinosaurs, must have been to produce brilliant displays of color. The discriminating dinosaur eye would have been well adapted to perceive the electric hues of a courtship signal or the scarlet flashes of a threat display. Feathers later acquired another function—flight—by the lineage that became birds, but they remained a key survival advantage for *Anzu* and other feathered, flightless dinosaurs.

So far we've been discussing predators, but some dinosaurs specialized in defense. *Ankylosaurus*, for example, had features that might leave a tank commander verklempt. Twenty feet long, thickly armored, and in possession of a devastating counterpunch, *Ankylosaurus* would have dissuaded all but the most maniacal aggressors. Its body proportions were reminiscent of an American football fullback. With relatively short legs and a thick torso, it held its thirteen tons about a low center of gravity, which made the squat plant eater difficult for enemies to topple. Osteoderms (armor plates) lined its back, not unlike shoulder pads. Its brain and cranial sensory organs were guarded by a thick helmet of bone. *Ankylosaurus* skulls show no sign of the airiness characteristic of theropod and sauropod crania. Instead, they possessed only the minimum complement of openings to the outside world, those necessary to sustain life: a mouth, eye orbits, and nose holes. Its head resembled a flattened turtle skull, an example of evolutionary convergence under similar selection pressures. The hip bones (ilia) of *Ankylosaurus* were tilted up, shielding its visceral mass, when hunkered down, from assault from above. Although an open hip socket is a defining characteristic of dinosaurs, and their ancestors possessed this trait, it was lost in *Ankylosaurus*, who evolved a closed, cup-like socket to receive the femur—one less point of entry for an ill-intended claw or a malicious tooth. If these daunting fortifications failed to dissuade would-be attackers, *Ankylosaurus* could launch a devastating counterstrike. Its 8-foot (2.5-meter) tail terminated in a terrible club, something akin to three bony volleyballs fused together into a medieval mace. And it could wield this weapon over a 100-degree arc while accelerating it to bone crushing speeds.[76] Any attempt to dine on *Ankylosaurus* meat would have been met with potentially lethal resistance.

Standing and fighting is one way to thwart predators. Running away is another. Hadrosaurs, duck-billed dinosaurs, were champions of flight. With teeth specialized for dismembering plants, no claws, no horns, no clubs, and no armor, fleeing danger was their only choice. The many species of hadrosaurs were probably North America's most common dinosaurs during the latter portion of the Cretaceous Period. They are certainly represented by the most numerous dinosaur fossils. One duckbill, *Edmontosaurus*, is sometimes found in bone beds containing thousands of individuals. A single deposit in eastern Wyoming contains as many as twenty-five thousand. In Alaska's Denali National Park, thousands of hadrosaur tracks are preserved in sediments laid down by an ancient river that once flowed across a gently sloping coastal plain. The wide range of track sizes shows that the region was home to multigenerational herds of duckbill dinosaurs. Like many large herbivores today, they must have lived gregariously within a social structure.[77]

From a scientific standpoint, hadrosaurs include the best-known dinosaurs, owing to the great number of specimens recovered and the extraordinary preservation of some. We know hadrosaurs from their bones, teeth, skin impressions, gut contents, eggs, embryos, and trackways. Even the traces of gut parasites of one duckbill were studied. And collagen, blood vessels, and blood cells were recovered from another.[78] While most dinosaur species are known only from a single partial skeleton, many hadrosaur species are known from multiple complete specimens.

Duckbills seem to have had a penchant for being caught up in unusual geological situations, the kind that result in extraordinary preservation. Many hadrosaur tracks, including most of the Denali prints, preserve scaly skin impressions. Some duckbills are even known from 3-D body fossils: the so-called mummy dinosaurs.

One such specimen has long been displayed at the American Museum of Natural History: a "mummified"

*Edmontosaurus*. Though I've seen it many times, the shock of being in a room with a fleshed-out dinosaur never diminishes. I recoil in amazement, and a little bit of disgust, each time I see it. The headless specimen lies spread-eagle on its back like a giant butterflied chicken about to be grilled. Its bones are ghoulishly draped with a thin veneer of mineralized skin. Its once huge gut is desiccated and collapsed in on itself, giving it a stricken and emaciated look. Dinosaur roadkill comes to mind.

The rules of extraordinary preservation that would yield a 3-D body fossil such as this are still being worked out. The term *mummy*, though, is inappropriate. True mummies are specimens for which the bulk of the soft tissues have been preserved intact rather than mineralized. Mummification results from desiccation, freezing, or pickling.[79] Once mummies are removed from the conditions that promoted mummification, decay commences. This is the case with freeze-dried mammoths trapped in the permafrost. As the planet warms, more and more of these seem to be emerging from the thawing tundra. These specimens are not mineralized, though, but remain fully organic. Think of them as "mammoth jerky," not fossils. Technically, we would call them subfossils. Mummies, whether natural or human-made, are no more than thousands of years old, whereas the youngest non-avian dinosaur fossils are at least sixty-six million years old. Thus, the dinosaurs preserved as 3-D body fossils are not true mummies. They are, however, mostly hadrosaurs. Why?

Hadrosaurs seem to have favored coastal-plain environments. Frequently their remains are found in coastal deposits and in meandering lowland stream deposits. It was once thought that hadrosaurs, such as *Hadrosaurus foulkii*, were particularly adapted for consuming aquatic plants. There is little evidence for this, however, and debate continues over whether they tended toward grazing (feeding off ground plants) or browsing (feeding off shrubs and trees). Whatever their diet, they did seem to have preferred lowlands and wetlands. And so do sediments.

Upland areas, such as mountains and hillsides, tend to shed sediment under relentless attack from an army of geological processes. That sediment works its way downslope, where it eventually settles into bottomlands and along coasts. Since dead bodies must first be entombed in sediment in order to transition from the biosphere to the geosphere, lowland species are at an advantage with respect to representation in the fossil record. In fact, for this reason, ancient alpine ecosystems are almost never preserved. Considering the long duration of the dinosaurs and their ubiquity across Mesozoic continents, I would be shocked if there weren't dinosaurs specifically adapted to alpine biomes. Yet we know nothing of them, and probably never will, because of preservational biases built into the fossil record.

Hadrosaurs, with their affinity for lowland environments and tendency to congregate in herds, seem to have had a better chance of entering the fossil record than any other type of dinosaur. With so many chances to become fossils, it is not surprising, based on probability, that they are represented by the most numerous and the best-preserved specimens.

A quick scan, though, of children's literature about dinosaurs or a quick internet search for dinosaur toys will reveal that duckbills are a group largely overlooked by the public. Lacking the scary claws and teeth of the theropods, the size of the sauropods, and the spectacular defenses of the armored dinosaurs and the horned dinosaurs, duckbills have a charisma problem. They were impressively speedy but probably not the fastest dinosaurs. That title goes to the ornithomimids, who zipped along like giant nightmarish ostriches. No, the duckbills' superpower is more subtle but no less effective. Hadrosaurs were champion chewers. I realize that super-efficient mastication of plant material probably doesn't warrant its own scene in *Jurassic Park*. But this superpower was the key to their success.

The skulls of duckbills are quite remarkable. In the front of their mouths, they had no teeth but a beak that acted like a set of garden sheers for clipping plants. They did not suffer for want of teeth, though. Along the sides of their mouths, hadrosaurs had complex stacks of teeth, bound together by ligaments, that formed dental batteries. Each battery consisted of rows of dead teeth at the grinding surface, backed up by stacks of growing teeth held in reserve below. Because the active teeth were dead, they could be painlessly worn down into little nubs before being replaced.[80] Some species had as many as three hundred teeth. Paleontologist Robert Reisz at the University of Toronto, who studies hadrosaur dentition, calls it "the most complex dental system ever."[81] What's more, their teeth occluded in an elaborate motion that maximized chewing efficiency. The skulls of hadrosaurs, and their kin, possessed an upper jaw (maxilla) that was capable of hinging outward at the conclusion of every bite.[82]

A leaf fancied by a hadrosaur would be nipped off by the beak and transferred to the dental batteries. There, between powerful jaws, it would be crushed, as if in an olive press, bursting the cell walls. Then, as the upper dental battery hinged outward, the leaf would be ground, as though put through a pepper mill. After a few chews, there would be very little structure remaining. With its cells walls torn asunder and its nutrients laid bare, the decimated leaf would slide down the esophagus into a copious brew of bacteria and acids, where the energy borne within and now unbound would be turned into more and more hadrosaur. This efficient system of converting plants into dinosaurs would power sprawling herds of duckbills, as they snipped, chomped, and ground their way across the fertile plains of the Cretaceous. While not spectacular in outward form, hadrosaurs are a special kind of amazing.

Every group of dinosaurs was remarkable in its own way, but surely the most majestic of all must have been the sauropods. Although there were some relatively diminutive forms, such as the elephant-sized *Saltasaurus*, many members of this group grew to nearly unimaginable proportions. The tallest, longest, and most massive animals to have ever walked upon the Earth were sauropods. In 2005, I discovered such a creature.

# 10 *Dreadnoughtus*

It was my second season prospecting for dinosaurs in the remote and desolate badlands of southern Patagonia. We had pitched camp along the banks of a roaring glacial river within view of the rugged, snowcapped Andes Mountains. The first new species of dinosaurs were found near London and Philadelphia by gentlemen paleontologists who could excavate fossils and be home for dinner. How incredibly convenient! With the low-hanging fruit picked decades ago, paleontologists now find themselves going to the farthest reaches of the globe to make new discoveries. This was my situation.

Getting to my field site from the United States required transiting through six airports. Once on the ground, in the picturesque Patagonian town of El Calafate, a rocky road lay ahead: a three-hour drive up Argentina's legendary Ruta 40, a dirt-road version of US Route 66, then another hour off-road, on a track I had plotted through the scrubland.

We arrived in January, the austral summer. But it never felt like summer. At that latitude, 50° S, Antarctica holds sway over the weather. Working in our parkas was not unusual, and on particularly frigid mornings, the drinking water inside my tent would be capped with a layer of ice. And, of course, anyone who's ever been to Patagonia knows that the wind never stops. During the previous year, I had provisioned cereal for the crew to eat for breakfast. Bad idea. Cereal blows off your spoon in Patagonia before you can eat it.

I had come to this particular spot on Earth because it possessed four essential qualities. It had rocks of the right age for someone looking for dinosaurs. They were terrestrial sedimentary rocks, which means that fossils could be preserved in them and dinosaurs were a possibility. The site sits in an arid environment, so I could count on erosion to push back the hillsides and expose fossils at the surface, And, lastly, since a German explorer recorded a few scraps of dinosaurs bone on the other side of the river in 1922, there had been only occasional and casual prospecting in the area, and no serious collecting. The chances of finding bone were near 100 percent, and, in my estimation, the chance of finding something new to science was high.

I picked my way up a rocky arroyo, walking by those Native American hand axes that archaeology student Brenda Gilio would discover two years later. Anticipation ran high. This was giant country. A big discovery could be around any corner.

After recording the locations of a dozen or so sites with bone fragments, I spotted a small patch of fossil bone exposed along the side of a steeply sloping mountain. My initial excitement was quite muted. When you're in the right geological situation, it's common to find bits of bone. But the vast majority of such discoveries consist of unusable fragments, or isolated bones, which yield limited information. Within an hour, we had uncovered a femur that stretched 6 feet 3 inches (191 c). The previous year, we had found a femur that was 7 foot 3 inches (222 c), but after days of digging, that bone proved to be isolated. So my reaction to seeing another giant femur poking out from the sandstone on the flanks of Cerro Fortaleza was guarded.

Soon, though, as we continued to dig, the fibula (lower leg bone) appeared right where it belonged, and then the tibia (shinbone) next to it. This site had promise. Three bones, essentially connected, is a very good sign. And the dinosaur we were uncovering was huge! We could see already that it was one of the biggest dinosaurs that had ever been found. But who was it? I knew that the other huge dinosaurs found in South America were from rocks millions of years older. This was most probably a new species, but leg bones are not especially diagnostic. There was no way to be sure unless more of its skeleton was preserved beneath us.

A second femur appeared, then some ribs, and then a line of tail vertebrae. By the end of the day, we had exposed ten bones. This was a good site. This was a good day. We had found what we were looking for. We hiked back to camp,

It took four austral summers in the whipping, frigid winds of Patagonia to free *Dreadnoughtus schrani* from the rocks. One hundred forty-five bones were recovered, which paint a picture of a beast eighty-five feet long, two-and-a-half stories at the shoulder, weighing an astounding sixty-five tons.

This gargantuan plant-eater dominated its ecosystem and had little to fear from even the largest predators, which it outweighed by nine times.

fished some cans of Quilmes beer out of our "refrigerator," a section of the river penned in by rocks, and celebrated.

Over the next several days, we uncovered bone after bone. But as we dug deeper, my buoyant mood turned to worry. I had never seen so many large bones in one place. Few have. And we were in an extremely remote location, with only our camping gear, shovels, pickaxes, and a small crew. As the expedition leader, my mind turned over and over on the logistical hurdles that lay ahead. Solving the problems associated with excavating, transporting, and studying this huge sauropod skeleton would occupy my thoughts and actions for the rest of that field season and for much of the decade to follow.

We dug for two months that year, in the cold, wind-blasted Patagonian summer, new bones appearing every day. We returned the next year, and the next, and the next. By the end of four field seasons, our hardened and weather-beaten crew of students and volunteers—augmented by the occasional visiting colleague—had recovered 145 bones from a gargantuan new species of dinosaur.

With countless permits in hand and an insurance policy on a dinosaur written by Lloyd's of London, we loaded sixteen tons of plaster-jacketed bones into a shiny orange Hamburg Süd shipping container, which we accompanied to the Argentinean port town of Río Gallegos. From there the fossils traveled by sea up the coast of South America, across the Caribbean, and along the Eastern Seaboard to the port of

Philadelphia, where they were off-loaded and trucked to my lab. They would reside in the United States for five years on a research loan granted by the government of Argentina. To speed the preparation of the fossils, which entails painstaking removal of the remaining rock and, sometimes, chemical stabilization of the bones, I split the haul into three portions, sending one to the Carnegie Museum of Natural History and one to the Academy of Natural Sciences, and keeping the remaining third in my lab. Several years in, we were able to reunite the bones in my lab and identify eight unique skeletal features that set this dinosaur apart from all others, as a never-before-seen species.[83]

Naming a new species is among the most satisfying rewards of paleontology but also, in my view, a solemn and weighty responsibility. In crafting a name for the new Patagonian titan, I had hoped to reflect and to honor the toughness and ferocity that this sauropod must have possessed. I have often fretted about the dopey way in which these giants have been presented to the public in artistic reconstructions, a theme that began with Charles R. Knight.

In 1897, the same year that Knight painted the inspired and insightful *Leaping Laelaps*, he also tackled the giant sauropods *Diplodocus* and *Brontosaurus*. (Yes, *Brontosaurus* is back! A 2015 study found it to be distinct from *Apatosaurus*, resurrecting the much beloved species.[84]) The smaller *Diplodocus* in Knight's painting is relegated to the background, and *Brontosaurus*, the largest-known dinosaur at the close of the nineteenth century, is featured. The painting is stunningly beautiful; one of my favorites. But like all of Knight's works, his art is as much a snapshot of his own time as it is a moment frozen in time from the annals of our ancient world. In it he depicts a *Brontosaurus* languidly occupying a lush and verdant swamp. Other members of its species can be seen, mostly submerged, buoying their great masses for a respite against gravity. Its feeding interrupted, the *Brontosaurus* gazes absently at the viewer with no more malice than could be mustered by a duck. "Gentle giants" seems to be the message. Lumbering titans that spent their days bathing in sunny pools, waiting politely for theropods to come along and take a bite out of them. If present-day herbivores are a guide, this could not have been further from the truth.

Big herbivores tend to be territorial, surly, and aggressive. Far more people have been injured in Yellowstone National Park by bison than by grizzly bears. In Africa, hippos are widely regarded as the most dangerous large animals. And in India, as humans expand into previously wild areas, human-elephant conflicts have become a major problem, with as many as three hundred people killed each year. Big herbivores are dangerous.

By orders of magnitude, our new sauropod dwarfed every land animal alive today. It was eighty-five feet (twenty-six meters) long, from snout to tail. It stood two and a half stories at the shoulder. And all fleshed out, in life, it would have weighed 65 tons (59.3 metric tonnes). That's the mass of thirteen bull African elephants. That's the mass of about nine *T. rex*. That's about ten tons heavier than a Boeing 737.

Can you imagine a 65-ton bull sauropod in the breeding season, defending a territory? That animal would have been incredibly dangerous, a menace to all around. And it would have had nothing to fear from any other creature in its environment. Considering this, I gave the new dinosaur the name *Dreadnoughtus*, meaning "fears nothing." The name is an allusion to the turn-of-the-last-century battleships, the dreadnoughts. At the time of their construction, they were essentially impervious to the preexisting technology of war and had nothing to fear from any other class of warship at sea. For the species name, I choose to honor the Philadelphia high-tech entrepreneur Adam Schran, who generously supported our work. Thus, *Dreadnoughtus schrani* became the name.

*Dreadnoughtus* is remarkable in many ways. It is the most massive land animal for which we can confidently calculate a weight. Giant sauropods did not spend their days half submerged in water, like half-stranded land whales. With their massive, columnar limbs, they were fully terrestrial creatures capable of walking overland and standing for extended periods of time. In fact, it is questionable whether the largest ones could lie down at all. They may have stood their ground around the clock. To support their great mass, the limb bones of huge sauropods were perfectly scaled to the structural demands of their lifelong war against gravity.

Most quadrupeds possess a fairly consistent inventory of limb bones inherited from their ancient amphibious forebearers. Although lower limbs may be greatly modified—horses, for example, have lost most of their foot bones and essentially walk on their middle digits—all quadrupedal animals have only a single bone in each upper limb: a humerus or a femur. Each of these bones must be sufficiently sized to bear the entire weight that rests upon them. Therefore, it is not surprising that in extensive studies of living quadrupeds, the minimum circumference of both the humerus and the femur corresponds closely with the weight of the animals measured. Applying the same method to the humerus and the femur of *Dreadnoughtus* yields a weight of about 65 tons (or 59.3 metric tonnes). There are other huge dinosaurs of similar size, such as *Argentinosaurus* and *Futalognkosaurus*, but their recovered skeletons lack the limb bones necessary for these calculations, so we can only speculate about the tremendous mass they must have possessed.

Moreover, *Dreadnoughtus* stands out for its completeness. Giant sauropods are often represented by fragmentary remains. Imagine an animal that is literally the size of a house. If that animal dies and falls over on a hard substrate, such as a floodplain, at that moment, very little of its body is in contact with the Earth. As a result, very little, if any, of its skeleton will have the opportunity to become buried by sediments before scavengers pull apart the remains or before the elements weather the bones to dust. Thus, many sauropod species are known from only a few bones. Another Patagonian giant,

*Puertasaurus*, for example, is known only from three bones: a neck vertebra, a partial back vertebra, and a tailbone fragment. Prior to the discovery of *Dreadnoughtus*, the most completely known supermassive dinosaur (those more than forty tons) was *Futalognkosaurus* from northern Patagonia. Excluding its head, 27 percent of the types of bones in its skeleton were recovered. Using the same metric, *Dreadnoughtus* is known by over 70 percent of its skeleton.[85] The remarkable completeness of *Dreadnoughtus* gives us a wonderful opportunity to study the anatomy, and therefore the biology, of one of the largest creatures to have ever walked on land.

How did dinosaurs like *Dreadnoughtus* grow so large? Efficiency. They ran high body temperatures, simply by virtue of their great mass. In fact, it seems that cooling, not heating, was their major thermoregulatory challenge. *Dreadnoughtus* has a thirty-foot tail and a forty-foot neck. That would have given it a huge amount of surface area per volume, making it essentially a giant radiator of heat. Additionally, its bird-like lungs were connected to large internal air sacs fore and aft of the lungs, providing great capacity to dump heat using its respiratory system.

With its long neck, *Dreadnoughtus* could stand in one place and clear out a huge patch of vegetation, taking in tens of thousands of calories while expending very few. In contrast to its brutish body, set upon feet that pressed into the earth like four heavy boulders, its lightly built head swung breezily about the leaves and fronds that were its life's obsession. Relative to its body, its head was minuscule—not much larger than a horse's head. If you've ever held a stack of books at arm's length, you can appreciate that putting a lot of weight at the end of a forty-foot lever (its neck) is a bad idea.

The small head of *Dreadnoughtus* was minimalistic, formed by thin bones that were weakly sutured together. While hadrosaurs epitomized dental complexity in dinosaurs, sauropods, by contrast, were austere and utilitarian—the midcentury modern of dentition versus the hadrosaurs' Gothic cathedrals. The teeth of *Dreadnoughtus* were simple pegs incapable of chewing; good only for ripping vegetation from the landscape. Leaves and fronds arrived whole and unmolested in their copious stomachs, which must have been huge, probably the size of a horse. Their never-chewed food was processed biochemically by a multitude of bacteria that lived generation after generation in the dark, dank recesses of their cavernous guts. There plants lingered and fermented for days, or maybe weeks, wending their way through a caustic digestive brew, in which microbial symbionts labored day and night, providing nourishment to the surrounding mountain of dinosaur flesh.

No doubt, *Dreadnoughtus* was an obsessive eater, to the exclusion, I think, of nearly everything else. An imaginary day planner for one might look something like this:

Monday—Eat, eat, eat, eat, eat, poop, eat, eat, eat, eat, eat, poop, eat, eat, eat, eat, eat, stop eating for a bit while it's dark.

Tuesday—see Monday. Wednesday . . . you get the idea. A few times a year, perhaps, replace a couple of those *eat*s with *sex*. Possibly throw in a pair of migrations, six months apart. That's about it. If you're going to pack on 130,000 pounds, there isn't time for much else.

Every dinosaur species deserves a book that would hold its lessons and reveal its secrets. Indeed, the current volume of information about dinosaurs is far too great to be contained within any single work. With an average of three new dinosaur species published each month,[86] it's hard to keep up. Even titles such as *The Dinosauria* and *The Complete Dinosaur* offer only abstracted views of the ocean of information about dinosaurs that washes across scientific journals and websites every day.

It's clear that we've only scratched the surface when it comes to the richness and breadth of dinosaur biodiversity. And dinosaurs represented only a fraction of the lush and bountiful ecosystems of the Mesozoic. Today biodiversity is even higher. Organisms have gotten better at partitioning themselves into more and more specialized niches, and the number of species now is perhaps twice what it was during the Cretaceous Period. But we're losing them fast—alarmingly fast. Today one out of three vertebrate species is threatened with extinction, and nearly half of invertebrate species are in peril. Here again, the dinosaurs can teach us from their rocky graves. Among the cardinal rules of evolution, one stands inviolable: no species lasts forever. Even the mighty dinosaurs, who dominated Earth's continents so thoroughly for so long, were incapable of maintaining their hegemony indefinitely. They had adapted to changes in climate and reconfigured continents. They took advantage of the arrival of new plants and coped with the extinction of others. They managed to spread themselves across the globe and insinuated themselves into every environment where there was air to breathe. But all this took time. If variation is the fuel of evolution, time—deep time—is the tracks on which she runs. After an epic reign, the monarchs of an Earth era had run out of time, and it all came crashing down.

# 11 Dinosaur Apocalypse

It ended in a flash. A moment predestined by a small event, millions of miles away. An appointment etched into the cosmic calendar perhaps before the first dinosaurs. A space rock nudged from its berth in the asteroid belt. A rock so old that the first rays of light to leave our star once shone upon it. Dislodged from its benign companions, it joined the ranks of the near-Earth objects: asteroids and comets that dive deep into the solar system with the power to menace our world. The smallest perturbation along its ancient path would have done the job. Whether by the vague gravity of a passing asteroid or by a mere encounter with an orbiting pebble, the rock was now set on a collision course with Earth. After a billion or so revolutions around the sun, there would be but one more. One last turn, and then the pretty blue planet would lie in its path. Beneath the swirling clouds of Earth, life abounded. It wasn't always this way.

The asteroid once shared the solar system with an Earth devoid of any living thing—a planet boiled and sterilized by insistent cosmic bombardment. When the worst of it ended, life grabbed hold and never let go. For three billion years, the space rock circled the sun, while nothing more complex than clumps of microbes agitated in our planet's primordial ocean.

It was out there, floating in space, when cells on Earth began to cooperate, building bodies and grouping themselves into tissues. It slowly arced through Cambrian skies when the major branches of the tree of life emerged from our common stock. When lobe-finned fish climbed out from the seas, it was there, tracing endless ellipses in the void. There, when the sauropsids split from their kin. There, when dinosaurs were new and could terrorize nothing bigger than a bug. There, in the sky, over every muddy *Spinosaurus*, every quarrelsome *Dryptosaurus*, every bloodthirsty *Anzu*. There, moving, always moving, silently above, unbeknownst to every unbudging *Ankylosaurus*, every flatulent *Hadrosaurus*, and every insatiable *Dreadnoughtus*. There, during the whole reign of the dinosaurs, a cosmic sword of Damocles, dangling, always, over their exquisitely evolved heads. There, waiting to drop, waiting to usher in a new age of Earth.

Speeding toward its rendezvous with history, the eight-and-a-half-mile-wide (14 km) asteroid was dwarfed by the shiny blue planet beneath it, eight thousand miles across. The size disparity would have appeared extreme, a speeding gnat heading for an elephant. Nothing more than a pinprick, right? The Earth is big, 197 million square miles (510 million $km^2$) across its surface. The impact blew a 125-mile-wide crater in the ocean crust beneath the Gulf of Mexico: a 12,000-square-mile (31,500 $km^2$) hole known as the Chicxulub crater. That's less than 0.01 percent of the Earth's surface—a pinprick. The planet was fine. There were a few days of seismic rumblings, and some volcanic fissures in India gurgled to life and spewed lava across the subcontinent,[87] but no appreciable damage to the planet as a whole.

It takes a lot to hurt a planet. The comedian George Carlin once framed our current environmental crisis this way: "The planet is fine . . . [it] will shake us off like a bad case of fleas. A surface nuisance." He was right, of course. When environmentalists talk of "saving the planet," they are really talking about saving life *on* the planet. The planet itself is all but guaranteed to exist for roughly another seven billion years. When the sun's stores of hydrogen are exhausted, about five billion years from now, our star will transform into a red giant. After another two and a half billion years, its surface will expand beyond the orbit of Earth. At some point, the Earth will fall to the sun, and the star stuff, from which our planet was forged, will become star stuff once more.

But by then, life will be long gone. About one and a half billion years from now, gradual solar brightening will cause a greenhouse effect that will spiral out of control.[88] Our oceans will vaporize, and the long-burning flame of life on Earth will dim, sputter for a bit, while extremophile bacteria linger on, and then go dark. When the last bacterium on Earth runs down, and its minuscule parcel of being spills through its ruptured membrane, the long experiment of life on this planet will conclude. This is why Michael Griffin, the former NASA administrator, said, "In the long run, a single-planet species will not survive." And, thus, as astronomer Carl Sagan put it,

"If our long-term survival is at stake, we have a basic responsibility to our species to venture to other worlds."[89]

Planets are tough. Life is fragile. The pinprick from space that barely altered the geology of Earth was devastating to the life on its surface. When we think of our planet, we tend to envision the things that surround us: the mountains, the oceans, lakes, forests, cities, the atmosphere. Bundled together, the Earth that we see, the Earth that we experience, amounts to next to nothing compared with the vastness of the entire planet. Think of a chicken egg. The thin shell that surrounds it represents about 2 percent of the thickness of the egg. By comparison, the atmosphere, roughly 60 miles high, is about 1.5 percent the thickness of the Earth. The biosphere is even thinner. The lowest point on Earth, the Challenger Deep, in the Mariana Trench, lies 35,462 feet below sea level. Mount Everest stands 29,029 feet above sea level. Except for aviators and their passengers, wafting microbes, and a few stupendously high-flying birds, life on our planet is essentially bounded by these topographic extremes. They form an envelope only twelve miles thick, 0.3 percent of the thickness of Earth. Shave down an eggshell by half, and then half again. Now take what's left and shave off another 40 percent. That's the thickness of the biosphere. It's not much. And it's easy to damage.

The force released by the impacting asteroid was the force of its momentum. It's a simple formula: $p=m^{*}v$. That is, the force of momentum (*p*) equals the object's mass (*m*) times the object's velocity (*v*). Bullets don't weigh much, but when shot from a gun, they can be devastating. A .30 caliber round, a bullet not quite a third of an inch across, weighing five one-hundredths of a pound, can bring down a moose that weighs three quarters of a ton. Take that same bullet and simply throw it at the moose, and you'll do nothing but annoy a moose, which is not a good thing to do. (Remember, herbivores are surly.) So it wasn't merely the mass of the asteroid that mattered. It was also its velocity. And the space rock heading for Earth was roughly the size of Manhattan and was traveling at forty thousand miles per hour—that's twenty-five times the speed of a bullet. It would pack a hell of a wallop, and the devastation to the living world would be apocalyptic.

A hadrosaur living on the coast of ancient New Jersey might have seen the asteroid streaking overhead, might have seen a flash in the southern sky. Startled for a moment, it listened and surveyed the scene around it. All quiet. Nothing amiss. A long breath in. No predator scent. Everything normal. Back to chewing. Look! There's a leaf! Nip, chomp, grind, grind. Nip, chomp, grind, grind.

While the hadrosaur chewed away contentedly, the dinosaurs within nine hundred miles of the impact were already dead. Vaporized or blown apart. For them, the Cretaceous had ended. Elsewhere in the world, dinosaurs were unaware of the tendrils of death that were reaching

The overarching lesson of natural history is that nothing lasts forever. Even the mightiest fall within the fullness of time. The long reign of the dinosaurs seemed as though it might last forever, but permanence in our world is an illusion. Since the flowering of complex life in the Cambrian Period, the tree of life has been severely pruned five times—five mass extinctions, the last of which put an end to the non-avian dinosaurs and 75% of species on Earth. For the dinosaurs, the fall was quick and violent. A hurtling asteroid, a moment of impact, and then earthquakes, flying debris, searing heat, tsunamis, and darkness—a calamity that cleared the way, as nature always eventually does, for something new.

out at that very moment to meet them. It was the last few minutes of the incomprehensibly long reign of the dinosaurs, and they carried on for a few moments more, obliviously doing dinosaur things.

At ground zero, the crust was splashing in breaking waves of molten rock across a crater larger than Massachusetts. Seismic shock waves, from a magnitude 10 earthquake, raced away in all directions at seventeen thousand miles per hour.

Eight minutes[90] had now passed since the hadrosaur saw the odd light in the sky. Eight minutes of chewing and farting and sniffing and stomping. Eight good hadrosaur minutes.

*Bang!* The ground convulsed with brutal violence. For the hadrosaur, and for most dinosaurs, this was the initial moment of terror; the moment when their world began to unravel. The seismic shaking would stagger some and knock others to the ground. Some big ones would die this way, lungs pierced by shattered ribs, organs burst, heads dashed upon the rocks. Closer to the impact, dinosaurs were tossed into the air by rolling waves of Earth. Trees would topple and hillsides fall.

The panicked hadrosaur ran wild-eyed, crashing through the standing timber and bounding over the fallen trees. Jagged branches tearing at its flesh. Bolting dinosaurs criss-crossing its path. A cacophony of birds and pterosaurs taking flight. And then the sky lit up, beautiful and strange, a red glow emanating from all directions.

The ejecta that splattered out from the crater had been thrown right through the sky. Among the debris were countless tiny blobs of molten rocks, which would cool into little glass balls called spherules. They were accompanied on their flight by fragments of rock and cubic miles of dust, bearing iridium from the asteroid itself.

The flying debris may have skirted along the upper boundary of the stratosphere, or it may have been launched clear into suborbital space. Either way, it would completely encircle the planet before settling back down to Earth.

At fourteen and a half minutes after impact, the sky began to fall over New Jersey, mostly fine dust and little glass spherules, with a few larger fragments. The deadliest moment had now arrived. Though the particles were small, it's not the size that is important, it's the collective gravitational energy stored in the particles. Some of the rock from the crater was vaporized on impact, but a large volume of it was flung out through the sky. Imagine how much energy would it take to lift cubic miles of rock to the top of the atmosphere. That's the amount of energy transferred from the asteroid into the flying debris. It's called potential gravitational energy. When the ejecta fell back to Earth, the potential turned into reality.

The collective mass, plummeting from the sky, had to dump all of that energy; had to balance the energy books. Some would have been dissipated through sound and some upon impact. But most would have been dissipated by friction with the atmosphere, releasing energy as infrared radiation and convective heat. Whether the mass of the ejecta is divided into large particles or motes of dust, the collective energy that must be dissipated is the same. As the bulk of the ejecta settled to Earth, unsheltered animals on the surface would have faced a searing blast of heat.

Sean Gulick, a geophysicist at the University of Texas at Austin, co-led a 2016 drilling expedition to the Chicxulub crater. There they collected core samples from the crater's rim and central peak, which will help clarify the magnitude of energy released by the blast. Atmospheric modeling has produced two leading scenarios for the heat pulse. Gulick described them to me as "toaster oven for a few hours versus pizza oven for a few minutes. "Either way," he added, "your dinosaur is dead."

Cooked, the hadrosaur's blistered body lay dead, with aftershocks, every few minutes, jiggling its still supple flesh. Two hours after impact, a wall of hot air, traveling at hurricane speed, whipped through the forest, knocking down a third of the trees. As the red glow began to fade, dust from the impact and soot from wildfires pulled a veil of black across the sky. The entire planet was plunged into darkness. Nighttime everywhere.

Tsunami waves propagated across the world's oceans. Hours after impact, a towering wall of water broke across the coast of New Jersey and pushed on for miles into the pitch darkness of the forest. Surviving animals would hear but not see the wave coming for them. After a devastating run-up, the roaring backwash would take with it broken trees, with rocks clinging to roots, scorched bodies of every description, and huge volumes of sand and mud liberated from the soil by earthquakes and uprooted trees and hurricane winds and landslides.

In the oceans, creatures outside of the blast zone were probably fine in the immediate aftermath. Some, in coastal areas, may have feasted on barbecued meat served up by the smoldering continents. A giant scavenger party broke out across the seas. Drifting bodies were gnawed by sharks. Carcasses on the sea floor were mobbed by worms and snails and sea stars.

But what was initially a boon turned grim quickly. As darkness fell upon the planet, marine plankton were deprived of the sunlight on which they fed. Starved, they faltered, and the foundation of the food web crumbled. Baseless, ocean ecosystems collapsed quickly. The lowest level of the food web, like the bottom of a pyramid, must be broad. This level is formed by the primary producers: the photosynthesizers, organisms that wondrously create food out of starlight. In the ocean, it is the phytoplankton that form the base, and the health and vitality of marine ecosystems rest upon them. As a rule of thumb, about 90 percent of energy is lost from one level of the feeding pyramid to the next. If you're going to be a thing, that eats things, that eats things, that eats plankton, there'd better be a hell of a lot of plankton. Otherwise there won't be much energy left for you at the top.

The hunger induced by the worldwide shutdown of photosynthesis was quickly transmitted up the feeding pyramid.

Soon the zooplankton, which fed on the phytoplankton, had little to eat. With plankton scarce, small consumers such as shrimp and tiny fish would have had a tough time. Most would make it, though, because their needs were modest. Larger carnivores would be hit harder. Many would go extinct. Some shark species died off, as did ammonoids, the coiled cephalopods whose seashells Mary Anning famously sold on the seashore. Predators at the top, like the mosasaurs and plesiosaurs, were doomed. The world could no longer satisfy such insatiable gluttony.

Bumping its way through waters choked with flotsam, the hadrosaur's bloated and rotting body drifted out to sea, a giant bobbing meat buoy, shedding flesh and dropping bones along the way. Spread over miles of seafloor, its remains and those of its ill-fated contemporaries were entombed by plumes of sediment that issued forth from the smoldering and blackened land. Its final sentence written, the Mesozoic Era, the epic trilogy of Earth history, drew to a close. Tucked in by a blanket of Earth and asteroid dust, the age of the dinosaurs was laid to rest.

A larger or faster asteroid might have wiped out complex life on our planet. A really big one might have sterilized it altogether. But as it was, about 25 percent of species survived. Crocodiles, turtles, and lizards suffered only moderate losses. Most sharks survived, although rays were hit hard. Nearly all boney fish species avoid extinction. But the flying reptiles (pterosaurs) and the huge marine reptiles (mosasaurs and plesiosaurs) would perish. On land, the scene was grim. The dinosaurs were devastated. Among them, only some birds would make it through. Mammals made it, obviously. This book was written by a mammal. But it was a close one. A recent study of North American fossil mammals found that in one region, 93 percent went extinct after the impact.[91] Doug Robertson, an emeritus fellow at the Cooperative Institute for Research in Environmental Sciences, and his colleagues hypothesized that the ability to shelter, either underground or in water, was the key to surviving the initial global thermal assault.[92]

For years, maybe a decade or more, darkness prevailed. The survivors languished in perpetual twilight under skies clouded with soot, dust, and droplets of sulfuric acid. Temperatures plummeted, and little grew during the long-impact winter. Thrown from a pizza oven into an empty refrigerator, it must have been too much for most. When it was over, 75 percent of species on Earth had vanished, in what looks like a geological instant. Iconic dinosaur species such as *T. rex* and *Triceratops* happened to be alive at the moment of impact. Following the carnage, the largest fully terrestrial creature on the planet was no larger than a modern cat. The age of the titans had ended, and the meek had inherited the Earth.

With dinosaurs gone, the mammals rose. Our shrew-like ancestors, and other mammals, had lived hidden in the dark recesses of the saurian world. Burrowing, and creeping, and foraging at night, their evolutionary possibilities were limited. When sunlight, and warmth, and blue skies returned to the world, opportunities abounded for the survivors. An explosion of evolution took mammalian species in all directions. One lineage became the largest animal ever: the blue whale. Although not much longer than a *Dreadnoughtus*, it possesses two or three times the mass. Out of the shadows, some land mammals would evolve into huge plant eaters and fierce predators. *Paraceratherium*, a hornless rhinoceros that lived around thirty million years ago, grew to twenty-six feet long and weighed fifteen tons. That's one and a half *T. rex*! Mammoths, some exceeding ten tons, spread across the Northern Hemisphere. Cave bears, saber-toothed cats, and dire wolves would terrorize Pleistocene[93] landscapes. Herds of bison, millions strong, would sweep across the Great Plains of North America, their hooves stamping on the graves of throngs of hadrosaurs that had thundered that way an era before. In South America, *Megatherium*, a four-ton ground sloth, would lazily bend tree trunks to its will. Its cousin *Doedicurus*, a giant armadillo, evolved into something like a mammalian version of *Ankylosaurus*. Tank-like, thickly armored, and brandishing a tail club, its description would warrant charges of plagiarism had these been fictional creatures.

Improbable as all the rest, our lineage evolved from our tiny, fuzzball Cretaceous ancestors. They did what no non-avian dinosaur could do. They found holes, and crevasses, and caves in which to shelter during the asteroid apocalypse, and they managed to eke out a living in the mangled landscape that was left. It wasn't long, only about ten million years, before a population of these survivors gave rise to the first primates. By about twenty-five to thirty million years ago, apes had diverged from monkeys. The apes we call humans, the hominins, parted ways with their chimpanzee kin around six million years ago. DNA analysis suggests that this was a drawn-out and apparently lascivious affair, with human-chimpanzee hybridization persisting for thousands or even millions of years. Many branches of humanity would sprout from this limb. It fact, it is unusual that we find ourselves alone today, the only humans on the planet. Just thirty thousand years ago, we shared this world with three other hominins: Neanderthals in Europe, Denisovans in Asia, and *Homo floresiensis*, the forty-inch-tall so-called hobbits from the Indonesian island of Flores. But as it is, we are the sole survivors of a primate lineage that can trace its ancestry back to a tenacious band of shrews that climbed from the ruins of the world's fifth mass extinction.

# 12 Why Dinosaurs Matter

*We are a way for the cosmos to know itself.*

—Carl Sagan

A few years ago, I bought a T-shirt for my wife at a local vineyard that bore its motto, "Everything Matters," a nod to its attention to the smallest details of viniculture. It's an idea that resonates with me—a hypothesis, really, that seems corroborated by innumerable, improbable turns of fate, recorded in the rock record. It's not that I'm sure that *everything* matters. I'm not. I'm not sure that it matters if I have a ham sandwich for lunch next Thursday. But on the other hand, I'm not sure that it doesn't. In chaos theory, this idea is known as sensitive dependence on initial conditions. It postulates that a tiny adjustment in the initial state of a deterministic, nonlinear system can result in huge differences in the end state. Popularly, it is known as the butterfly effect, from a metaphor something like this:

Hurricanes spawn from tropical storms, which form from tropical depressions, which spin into being from low-pressure cells, which grow from microdisturbances in the atmosphere. Thus, it is conceivable that a micro-low-pressure cell, spinning off the edge of a butterfly wing flitting through the Amazon rainforest, could result in a megahurricane striking the coast of the United States three weeks later.

The origins of natural phenomena can sometimes be determined using historical data. Scientists do a lot of this. A storm can be traced back to its origin, an earthquake to its epicenter, an epidemic to patient zero. Through the fossil record, whales can be followed back to wolf-like creatures on the ancient shores of Pakistan. And our species can be traced to the lakeshores and volcanic beds of the Olduvai Gorge, in Tanzania, that record the emergence of the human lineage. The retrospective approach can be difficult, but it works. Certainly much of the past has been lost to time, but scientists have grown amazingly adept at pulling bits of it back from the abyss. To a limited but increasing degree, the past is knowable.

Prospection, peering into the future, making predictions, is much more difficult. In fact, for complex collections of non-linear phenomena, we're terrible at it. Science is so poor at predicting the future of complex systems that is has yet to drive out the charlatans, quacks, and frauds that pretend to have the ability to do so. Fortune-tellers, astrologers, psychics, and other scam artists prey upon the common desire of people to understand the complexities that lie ahead. I was recently at a restaurant in Puebla, Mexico, that featured a table-side parrot that would select your fortune from a collection of predictions written on little tabs of paper. Such is our inability to peer into the future that, for some, taking advice from the parrot seems like as good a course as any.

To be clear, there are countless well-constrained phenomena that science excels at predicting, often with astounding accuracy, such as radioactive decay, planetary motions, tides, chemical reactions, the physics of light, aspects of climate change, and certain biological phenomena. When the NASA New Horizons probe rendezvoused with Pluto in 2015, after a journey of nearly ten years and three billion miles, it arrived at the planet within eighty-seven seconds of the time predicted at its launch. It was an extraordinary feat, a triumph of science and engineering. However, if I were to ask an equivalent group of experts to predict the color of the shirt of the person standing in front of me at Starbucks tomorrow morning, they would have no more than a random chance of being correct. Nor would they be particularly successful at predicting the outcome of the next Super Bowl, the weather one year from now, or whether they will be happy with their jobs in five years. The future, with respect to complex systems, is largely opaque.

Consider the mighty Missouri River. At its mouth, near Saint Louis, it is five hundred yards wide and discharges thousands of gallons of water per second into its confluence with the Mississippi River. Seeing the end, it's easy to imagine that the Missouri River travels a great distance and drains a huge part of the continent. In retrospect, it's easy to see that this river is a big deal.

But go to the head of the Missouri River—I once did—and its future is not apparent. After climbing to within a hundred yards of the Continental Divide, I found its beginnings and straddled its source. There the mighty Missouri is nothing more than a gurgle of water that issues forth from beneath a bolder, in a pasture, high in the Bitterroot Mountains. The tiny spring next to it travels a few hundred yards and ends in a small pond. The two streams, they look identical, but one is an anonymous trickle of water, and the other is the Missouri River. From a prospective viewpoint, they look the same—equally unimpressive. The Missouri River, at its head, does not look like anything special.

Now journey back to the Cretaceous Period and look at our tiny, skittish ancestors nervously navigating the dinosaur planet. With no *a priori* knowledge, would you select them as the winners in this world, as the one species that would spread across the globe, dominate every ecosystem, command fire, develop speech, invent art and science and engineering, and even venture forth to other worlds? Would you overlook mighty *T. rex*, pass by *Triceratops*, and scoff at thundering herds of hadrosaurs? Would you look at our wide-eyed ancestors, trembling in a hollow log, hoping never to be noticed, and think, *these are the ones.* These are the creatures that will navigate the seas, fight battles in the sky, and harness the atom. These are the creatures from whom this planet's

Einstein will emerge. These are the creatures that will walk on the Moon. These are the creatures that will study and name and grow to love the very dinosaurs that they dread. I think not. Lacking an omniscient Mexican restaurant parrot, you would never predict the future that awaits these tiny shrews. Nor would you anticipate the sudden, cataclysmic demise of the dinosaurs. Standing on the Continental Divide, one rivulet of water is as good as the next. It's all so contingent.

The dinosaurs were thriving the day before the asteroid hit. After 165 million years, why not another 66 million more? If the asteroid had missed, the pivotal, calamitous day that wiped out the dinosaurs and 75 percent of species on the planet would have been just another day. Just another day among the 63 billion days already enjoyed by the dinosaurs. But over geological time, improbable, nearly impossible events do occur. Along the path from our wormy Cambrian ancestors to primates dressed in suits, innumerable forks in the road brought us to this very particular reality. But would it happen this way again? It's nearly impossible.

If you made a thousand more Earths, in a thousand more solar systems, and let them run, you would always get a different result. No doubt, those worlds would be both amazing and amazingly improbable, but they would not be our world and they would not have our history. There are an infinite number of histories that we could have had, but we only get one. And, from a human perspective, we got an especially good one—the one that led to us.

Of course, I'm particularly grateful that there are humans in this world. I married one, and with the exception of a cat and a few fish, all my friends are humans. But beyond our self-interest, how great is it that we live in a world that had, and has, dinosaurs? By degrees, nearly everyone shares feelings of amazement and curiosity for these iconic beasts.

So ingrained are dinosaurs in our conception of the past that they transcend their biological definition and serve as a cultural construct for our fascination with the ancient. They stand in for the common awe and wonder that we feel when contemplating the past. When the coelacanth turned up alive in 1938, it was dubbed the "dinosaur fish." There's nothing dinosaur about it, of course, but the word connotes the long-lost, the ancient. Another Lazarus species, the Wollemi pine, thought extinct since the Mesozoic, was discovered alive in Australia in 1994. The press called it the "dinosaur tree." I doubt if anyone could truly be that taxonomically befuddled. Rather, the reporters used *dinosaur* as shorthand for the ancient and for something thought to be extinct. Perhaps it is their size, their power, their ferocity, or their charisma, but it is dinosaurs, more than any other ancient life-form, that have become the vessel for the emotional and revelatory power of the ancient past. When *Jurassic World* opened in the summer of 2015, smashing box-office records by raking in over a half billion dollars in a weekend, it was the latest installment of the dinomania that began with Benjamin Waterhouse Hawkins' Crystal Palace sculptures.

Clearly, dinosaurs strike a resonant chord within our imaginations. From where we humans stand today, they are our looking glass. They provide a view that is mind altering; a view that leads us to an elementary yet profound conclusion when we ask the question "Why study the ancient past?" For me, the answer is: because it gives us perspective and humility. The present is nothing but a fleeting instant. Gone before you can think of it. It's only the past that gives context to our world. And it's the past that gives us foresight. Dinosaurs matter because the future matters.

The hubris of humankind falters when our brief history is set upon the vast backdrop of deep time. Our self-importance wanes when we view our species as the occupants of but a single tiny twig among the multitudes that form the prodigious crown of the spectacular tree of life. And the rock record shows us that we are not inevitable beings; that we are not the recipients of an evolutionary proclamation of Manifest Destiny. We got lucky. And being lucky is a great feeling, but it should lead, I think, to gratitude, not pride.

The dinosaurs, the last great rulers of the land, died in the world's fifth mass extinction. They didn't see it coming, and they didn't have a choice. When the wreckage cleared, our meek and long-oppressed ancestors crept from their hidey-holes, ready to receive their inheritance: the next great era of Earth.

Now we are the rulers of the land and the sea. Meek no more, our species is propagating an environmental disaster of geological proportions that is so broad and so severe, it can rightly be called the sixth extinction. We are warming the planet, melting the glaciers, and raising the seas. The carbon dioxide that we pump into the atmosphere reacts with seawater and acidifies the oceans, killing the coral reefs. We're cutting down rainforests, filling in wetlands, and thawing the tundra. We're despoiling our environment with pesticides, heavy metals, and a witches' brew of flushed pharmaceuticals. A recent study[94] found the current rate of extinction to be a thousand times above the natural rate. We are the asteroid! And it's worth remembering that it wasn't only the non-avian dinosaurs and three-quarters of life on Earth that perished in the fifth extinction; the asteroid was destroyed as well.

The abrupt fall of the dinosaurs shows us that nature's order is susceptible to perturbations, and disruptions, and turnovers. Reigns end, even those stable enough to persist for tens of millions of years. The fossil record instructs us that our place on this planet is both precarious and potentially fleeting.

But we are not dinosaurs. Unlike the dinosaurs, we can see it coming, and we can do something about it. The science fiction writer Larry Niven once quipped, "The dinosaurs went extinct because they didn't have a space program."[95] He's right. But we're not there yet, either. Neither do we have the ability to deflect an asteroid, in the near term, nor can we leave the Earth en masse. For us, there is no Planet B. We need to, as Carl Sagan put it, "preserve and cherish the pale blue dot, the only home we've ever known."

Dinosaurs matter, because the future matters. Among extinct creatures, they are not special in this regard. Every rock and every fossil has a tale to tell. Combined, they tell the story of our planet and are our guideposts into the future. Today, a sixth extinction is unfolding before our eyes. We are in the midst of it, but worse, we are the cause of it. We have become the asteroid of our age. The dinosaurs had no choice and played no part beyond dying in the unraveling of their world. This time, it's different. We can see it coming and we do have a choice. Armed with our knowledge of the past, can we rise to the challenge and secure our future?

There is no alternative. In the face of a bleak situation, we have to press on. The tireless warrior for climate responsibility, former Vice President Al Gore, cautions that there is no time to despair: "We have to win this struggle, and we will win it; the only question is how fast we win. But more damage is baked into the climate system every day, so it's a race against time."[96]

Perhaps the dinosaurs' long-lasting record of success is reason for optimism. If they persisted through so many changes, maybe we can too. But we have to act, and we have to act fast. Let's not be the asteroid. Maybe we can be like the dinosaurs, instead; the adaptable champions of an era.

## ACKNOWLEDGMENTS

I would like to express my gratitude to the following scholars for sharing their time and knowledge as I gathered information for this book: Dave Goldberg for his thoughts on Albert Einstein, and Alison Moyer for the clarity she shed on dinosaur feathers, both of Drexel University; Michael Habib, of the University of Southern California, for his insights into the functional anatomy of *T. rex*; Sean Gulick, of the University of Texas at Austin, for his enlightening discussion of the Chicxulub impact and its aftermath; Matthew Lamanna, of the Pittsburgh's Carnegie Museum of Natural History, for his comments on Anzu; and Harold Connolly, of Rowan University, and Alan Stern, of the Southwest Research Institute, who answered my questions regarding asteroid origins and behaviors.

I'm grateful to my wife, Jean, and son, Rudyard, who indulged me as I wrote portions of this book over weekends, evenings, and during family vacations in Maine and Colorado. Jean helped me collect my thoughts on this project and provided insightful commentary, as always, on the manuscript.

I am indebted to the passionate people of the TED organization, who work hard to make this planet a better and more interesting place. And I am grateful for the stylish artwork of Mike Lemanski, who illustrated key concepts in this book in a fresh and thought-provoking way.

Finally, I would like to thank my wonderfully kind and patient editor, Michelle Quint, who saw this book within me and encouraged me to write it, and whose keen sense of narrative arc, rhythm, and flow enhanced this work enormously.

## NOTES

1 Merriam-Webster Dictionary online, accessed May 9, 2016, www.merriam-webster.com.

2 Cambridge Dictionary online, accessed May 9, 2016, www.dictionary.cambridge.org.

3 Oxford English Dictionary online, accessed May 9, 2016, www.oed.com.

4 Matthew J. Dowd, "5 Takeaways from Indiana and the Path Ahead for Donald Trump and Hillary Clinton," Washington Wire (blog), *Wall Street Journal* online, last modified May 3, 2016, http://blogs.wsj.com/washwire/2016/05/03/5-takeaways-from-indiana-and-the-path-ahead-for-donald-trump-and-hillary-clinton.

5 Mesozoic Era (252–66 Ma). (Ma=millions of years ago), contains the Triassic, Jurassic, and Cretaceous periods. Sometimes referred to as "The Age of Reptiles." Cf. Geological Society of America Time Scale: Walker, J. D., J. W. Geissman, S. A. Bowring, and L. E. Babcock, (Compilers), "GSA Geologic Time Scale (v. 4.0)," www.geosociety.org/gsa/timescale/gts2012-commentary.aspx.

6 John Culhane, *Walt Disney's Fantasia,* reissued ed. (New York: Harry N. Abrams, 1999), 126.

7 Ibid., 123.

8 Cretaceous Period (145–66 Ma), in the Mesozoic Era, cf. GSA Time Scale.

9 Alvarez, L. W., W. Alvarez, F. Asaro, and H. V. Michel. "Extraterrestrial cause for the Cretaceous-Tertiary extinction: Experiment and theory," *Science*, v.208, pp. 1095–1108.

10 Usage note: In this book, for the sake of readability, I make sparing use of the cumbersome terms ***avian dinosaur*** and ***non-avian dinosaur.*** Instead, where it makes sense, I simply use the term ***dinosaur*** in common parlance, meaning non-avian dinosaur. In instances where the context makes it obvious, I use ***dinosaur*** to refer to all dinosaurs, avian- and non-avian. Where necessary, I distinguish between the two types.

11 Erin Wayman, "Five Early Primates You Should Know," Smithsonian.com, last modified October 31, 2012, www.smithsonianmag.com/science-nature/five-early-primates-you-should-know-102122862/#lfCf8ZSEhy504M8g.99.

12 Young, Nathan M., Terence D. Capellini, Neil T. Roach, and Zeresenay Alemseged. "Fossil hominin shoulders support an African ape-like last common ancestor of humans and chimpanzees." *Proceedings of the National Academy of Sciences* 112, no. 38 (2015): 11829–11834.

13 McDougall, Ian, Francis H. Brown, and John G. Fleagle. "Stratigraphic placement and age of modern humans from Kibish, Ethiopia." *Nature* 433, no. 7027 (2005): 733–736.

14 Gore, R., "The rise of mammals." *National geographic* 203, no. 4 (2003): 2.

15 Add a healthy plus-minus here for uncertainty in the geological record.

16 San Antonio, James D., Mary H. Schweitzer, Shane T. Jensen, Raghu Kalluri, Michael Buckley, and Joseph P. R. O. Orgel. "Dinosaur peptides suggest mechanisms of protein survival." *PLOS ONE* 6, no. 6 (2011): e20381.

17 Barrowclough, George F., Joel Cracraft, John Klicka, and Robert M. Zink. "How many kinds of birds are there and why does it matter?" *PLOS ONE* 11, no. 11 (2016): e0166307.

18 Jurassic Period (201–145 Ma), in the Mesozoic Era, c.f. GSA Time Scale.

19 Cambrian Period (541–485 Ma), in the Paleozoic Era, cf. GSA Time Scale.

20 As this book was in production, a paper was published in which the authors suggest a restructuring of the dinosaur phylogenetic tree, moving the theropod clade to the Ornithischian limb, and supplanting the name Ornithischia with "Ornithoscelida." At the time of this writing, a number of other research groups are endeavoring to test this idea (pers. comm., P. M. Barrett), so, in this book, it's too early to weigh in on this new hypothesis. Surely, though, it will take compelling evidence to alter the basic structure of the dinosaur tree, which has been agreed upon, in its broadest terms, for the past 130 years. Cf.: Baron, M. G., D. B. Norman, and P. M. Barrett. "A new hypothesis of dinosaur relationships and early dinosaur evolution." *Nature* 543, no. 7646 (2017): 501–506.

21 Prior to the Phanerozoic Eon, which began 541 million years ago, organisms lacked hard parts, such as shells, chitinous exoskeletons, bones, and teeth.

22 Peter Tyson, "Moment of Discovery," *NOVA* online, accessed August 26, 2016, www.pbs.org/wgbh/nova/fish/letters.html.

23 This point is illustrated beautifully in Neil Shubin's compelling book *Your Inner Fish: A Journey into the 3.5-Billion-Year History of the Human Body* (New York: Pantheon Books, 2008).

24 Archean Eon (4,000–2,500 Ma), cf. GSA Time Scale.

25 Devonian Period (419–359 Ma), in the Paleozoic Era, c.f. GSA Time Scale.

26 Carboniferous Period (359–299 Ma), in the Paleozoic Era, cf. GSA Time Scale.

27 Or the first mammaliaforms, depending on which definition is followed.

28 Adrienne Mayor, *The First Fossil Hunters: Paleontology in Greek and Roman Times* (Princeton, NJ: Princeton University Press, 2000).

29 Alan Cutler, *The Seashell on the Mountaintop: A Story of Science, Sainthood, and the Humble Genius Who Discovered a New History of the Earth* (New York: Dutton, 2003).

30 Hutton, James, "Theory of the Earth; or an Investigation of the

Laws observable in the Composition, Dissolution, and Restoration of Land upon the Globe," *Earth and Environmental Science Transactions of The Royal Society of Edinburgh* 1, no. 2 (1788): 209–304.

31 James Hutton, *Theory of the Earth, with Proofs and Illustrations, in Four Parts*, vols. 1 and 2 (Edinburgh: William Creech, 1795). Parts 3 and 4 were never published.

32 Hutton studied and published on numerous sites. Of them, Siccar Point is his most famous, though he died before he could describe this outcrop in his planned third volume.

33 Hutton, *Theory of the Earth*.

34 John Playfair published a concise and accessible book outlining Hutton's theory: *Illustrations of the Huttonian Theory of the Earth* (Edinburgh: William Creech; London: Cadell and Davies, 1802).

35 Robert Jameson published *Elements of Geognosy: The Wernerian Theory of the Neptunian Origin of Rocks* (Edinburgh: Bell and Bradfute, Guthrie and Tait, and William Blackwood; London: Longman, Hurst, Rees and Orme, 1808.) George Cuvier published his own theory of the Earth, based on six Great Deluges: *Recherches sur les ossements fossiles de quadrupèdes* (Paris, 1812), which became in translation *Essay on the Theory of the Earth*, trans. Robert Kerr, with notes by Robert Jameson, 2nd Ed. (Edinburgh: William Blackwood; London: T. Cadell, Strand, W. Blackwood, 1815).

36 Dennis R. Dean, *Gideon Mantell and the Discovery of Dinosaurs*, (Cambridge, UK: Cambridge University Press, 1999), 71.

37 Mantell, Gideon, "Notice on the Iguanodon, a Newly Discovered Fossil Reptile, from the Sandstone of Tilgate Forest, in Sussex," *Philosophical Transactions of the Royal Society of London* v. 115, (1825), 179–86.

38 Initially, the voyage was slated to map the coast of South America, but it was extended en route to circumnavigate the globe.

39 David Quammen, *The Reluctant Mr. Darwin: An Intimate Portrait of Charles Darwin and the Making of His Theory of Evolution* (New York: W. W. Norton & Company, 2006).

40 Jack Repcheck, *The Man Who Found Time: James Hutton and the Discovery of the Earth's Antiquity*. (Cambridge, Mass.: Perseus Book Group, 2009).

41 Stoddart, D., "Darwin, Lyell, and the Geological Significance of Coral Reefs," *The British Journal for the History of Science*, 9, no. 2 (1976), 199–218.

42 Charles Darwin, *Journal of Researches into the Geology and Natural History of the Various Countries Visited by H.M.S. Beagle, Under the Command of Captain FitzRoy, R.N., from 1832 to 1836* (London: Henry Colburn, 1839), a.k.a. *Voyage of the Beagle*, January 16, 1832 entry. (Also cf. Darwin Online: www .darwin-online.org.uk).

43 Ibid, January 16, 1832.

44 Ibid.

45 Ibid.

46 Edward Larson, *Evolution: The Remarkable History of a Scientific Theory*, (New York: Random House, 2004).

47 Charles Robert Darwin to William Darwin Fox, August 9–12, 1835, Darwin Correspondence Project, Cambridge University, letter number DCP-LETT-282, www.darwinproject.ac.uk/letter/DCP-LETT-282.xml.

48 Daniel C. Dennett, *Darwin's Dangerous Idea: Evolution and the Meaning of Life* (New York: Touchstone, 1995).

49 Repcheck, *The Man Who Found Time*.

50 Dennis R. Dean, *Gideon Mantell*.

51 Valerie Appleby, "Ladies with Hammers," *New Scientist* 84, no. 1183 (November 29, 1979): 714.

52 Hugh S. Torrens, "Politics and Paleontology: Richard Owen and the Invention of Dinosaurs," chap. 2 in *The Complete Dinosaur*, ed. Michael K. Brett-Surman, Thomas R. Holtz, and James O. Farlow (Bloomington: Indiana University Press, 2012), 25–44.

53 Owen's osteological definition of dinosaurs is no longer valid, and the modern definition relies on a different, expanded list of characters. He was, however, correct that Dinosauria represents a distinct clade and that *Megalosaurus*, *Iguanodon*, and *Hylaeosaurus* are rightful members of it.

54 Richard Owen, *Report on British Fossil Reptiles, Part II*, in the *Report of the British Association for the Advancement of Science for 1841* (London: Richard and John E. Taylor, 1841).

55 Cf. illustration of *Megalosaurus* from Richard Owen, illustrated by Benjamin Waterhouse Hawkins, *Geology and Inhabitants of the ancient World*, vol. 8. Crystal Palace Library, (London: Bradbury and Evans, 1854), 20.

56 *Megalosaurus* is from the Jurassic Period, while *Iguanodon* and *Hylaeosaurus* lived in the Cretaceous age.

57 Soraya de Chadarevian and Nick Hopwood, eds., *Models: The Third Dimension of Science* (Stanford, CA: Stanford University Press, 2004).

58 William B. Gallagher, *When Dinosaurs Roamed New Jersey* (New Brunswick, NJ: Rutgers University Press, 1997).

59 Ibid.

60 "August 21st," *Proceedings of the Academy of Natural Sciences of Philadelphia* 18 (1866): 275–79, www.jstor.org/stable/4059697.

61 Except for birds, no dinosaurs flew. The flying reptiles of the Mesozoic were pterosaurs, which are closely related to dinosaurs, but are not dinosaurs.

62 All non-avian dinosaurs lived on land. There is evidence, though, from swim-tracks, that they could swim, if necessary, like most animals. One non-avian dinosaur, *Spinosaurus*, has been hypothesized to be amphibious. Many birds, such as penguins, puffins, and cormorants, are excellent swimmers.

63 Brendan Maher, "The Theatre: Bringing the Past to Life," *Nature* 449, no. 7161 (September 27, 2007): 395–96.

64 Stevens, K. A., "Binocular Vision in Theropod Dinosaurs," *Journal of Vertebrate Paleontology* 26, no. 2 (June 2006): 321–30.

65 Richard A. Lovett, "T. Rex, Other Big Dinosaurs Could Swim, New Evidence Suggests," *National Geographic News*, last modified May 29, 2007, http://news.nationalgeographic.com/news/2007/05/070529-dino-swim.html.

66 Garm, A. and D. E. Nilsson, "Visual Navigation in Starfish: First Evidence for the Use of Vision and Eyes in Starfish," *Proceedings of the Royal Society B: Biological Sciences* 281, no. 1777 (February 22, 2014).

67 "Phenotype" refers to the sum total of an organism's observable traits, which results from the interaction of the genotype with the environment, whereas the genotype, alone, simply refers to heritable genetic identity.

68 M. Damian, R. Softley, and E. J. Warrant, "The Energetic Cost of Vision and the Evolution of Eyeless Mexican Cavefish," *Science Advances* 1, no. 8 (September 11, 2015): doi:10.1126/sciadv.1500363.

69 Brian Switek, "Stop Making Fun of Tyrannosaurs' Tiny Arms," Smithsonian.com, last modified March 31, 2016, www.smithsonianmag.com/science-nature/stop-making-fun-tyrannosaurs-tiny-arms-180958615.

70 Switek, "Stop Making Fun."

71 Marge Piercy, *Circles on the Water: Selected Poems of Marge Piercy* (New York: Knopf, 1982).

72 Kenneth Chang, "A Lost-and-Found Nomad Helps Solve the Mystery of a Swimming Dinosaur," *New York Times* online, September 11, 2014, www.nytimes.com/2014/09/12/science/a-nomads-find-helps-solve-the-mystery-of-the-spinosaurus.html.

73 Ibid.

74 Christine Dell'Amore, "New 'Chicken from Hell' Dinosaur Discovered," *National Geographic* online, last modified March 19, 2014, http://news.nationalgeographic.com/news/2014/03/140319-dinosaurs-feathers-animals-science-new-species; Lamanna, M. C., H-D. Sues, E. R. Schachner, and T. R. Lyson, "A new large-bodied oviraptorosaurian theropod dinosaur from the latest Cretaceous of western North America," *PLOS ONE* 9, no. 3 (2014): e92022.

75 Koschowitz, M-C., C. Fischer, and M. Sander, "Evolution: Beyond the Rainbow," *Science* 346, no. 6208 (October 24, 2014): 416–18.

76 Arbour, V. M., "Estimating impact forces of tail club strikes by ankylosaurid dinosaurs." *PLOS ONE* 4, no. 8 (2009): e6738.

77 Fiorillo, A. R., S. T. Hasiotis, and Y. Kobayashi, "Herd structure in Late Cretaceous polar dinosaurs: A remarkable new dinosaur tracksite, Denali National Park, Alaska, USA," *Geology* 42, no. 8 (2014): 719-722.

78 Schweitzer, M. H., W. Zheng, C. L. Organ, R. Avci, Z. Suo, L. M. Freimark, V. S. Lebleu et al., "Biomolecular characterization and protein sequences of the Campanian hadrosaur B. Canadensis," *Science* 324, no. 5927 (2009): 626-631; Tweet, J., K. Chin, and A. A. Ekdale, "Trace fossils of possible parasites inside the gut contents of a hadrosaurid dinosaur, Upper Cretaceous Judith River Formation, Montana," *Journal of Paleontology* 90, no. 2 (2016): 279-287.

79 Briggs, D. E. G., "The role of decay and mineralization in the preservation of soft-bodied fossils," *Annual Review of Earth and Planetary Sciences* 31, no. 1 (2003): 275-301.

80 LeBlanc, `A. R. H., R. R. Reisz, D.C. Evans and A. M. Bailleul, "Ontogeny Reveals Function and Evolution of the Hadrosaurid Dinosaur Dental Battery," *BMC Evolutionary Biology*, 2016, doi:10.1186/s12862-016-0721-1.

81 Elaine Smith, "With 300 Teeth, Duck-Billed Dinosaurs Would Have Been Dentist's Dream," Phys.org, last modified August 16, 2016, http://phys.org/news/2016-08-teeth-duck-billed-dinosaurs-dentist.html.

82 Nabavizadeh, A., "Hadrosauroid Jaw Mechanics and the Functional Significance of the Predentary Bone," *Journal of Vertebrate Paleontology* 31 (January 2014): 467–82.

83 Lacovara, K. J., M. C. Lamanna, L. M. Ibiricu, J. C. Poole, E. R. Schroeter, P. V. Ullmann, K. K. Voegele et al., "A gigantic, exceptionally complete titanosaurian sauropod dinosaur from southern Patagonia, Argentina," *Scientific Reports* 4 (2014): 6196.

84 Tschopp, E., O. Mateus, and R. B. J Benson, "A specimen-level phylogenetic analysis and taxonomic revision of Diplodocidae (Dinosauria, Sauropoda)," *PeerJ* 3 (2015): e857.

85 The heads of giant sauropods are so poorly known and are recovered so infrequently that we do not include them in our calculations.

86 Nick Longrich, "Why Were There So Many Dinosaur Species?," Phys.org, last modified December 13, 2016, https://phys.org/news/2016-12-dinosaur-species.html.

87 Richards, M. A., W. Alvarez, S. Self, L. Karlstrom, P. R. Renne, M.Manga, C. J. Sprain, J. Smit, L. Vanderkluysen, and S. A. Gibson, "Triggering of the largest Deccan eruptions by the

Chicxulub impact," *Geological Society of America Bulletin* 127, no. 11–12 (2015): 1507–1520.

88 Wolf, E. T., and O. B. Toon, "Delayed Onset of Runaway and Moist Greenhouse Climates for Earth," *Geophysical Research Letters* 41, no. 1 (January 16, 2014): 167–72.

89 Carl Sagan, *Pale Blue Dot: A Vision of the Human Future in Space* (New York: Random House, 1994).

90 The timing of events in this section is based on calculations using the Earth Impact Effects Program, by Robert Marcus, H. Jay Melosh, and Gareth Collins, accessed December 7, 2016, http://impact.ese.ic.ac.uk/ImpactEffects/.

91 Longrich, N. R., J. Sciberras, and M. Wills, "Severe Extinction and Rapid Recovery of Mammals Across the Cretaceous-Palaeogene Boundary, and the Effects of Rarity on Patterns of Extinction and Recovery," *Journal of Evolutionary Biology* 29, no. 8 (2016): 1495–1512.

92 Robertson, D. S., M. C. McKenna, O. B. Toon, S. Hope, and J. A. Lillegraven, "Survival in the first hours of the Cenozoic," *Geological Society of America Bulletin* 116, no. 5–6 (2004): 760–768.

93 Pleistocene Epoch (2.6 Ma–11,700 years ago), sometimes referred to as the "Ice Age," cf. GSA Time Scale.

94 Pimm, S. L., C. N. Jenkins, R. Abell, T. M. Brooks, J. L. Gittleman, L. N. Joppa, P. H. Raven, C. M. Roberts, and J. O. Sexton, "The biodiversity of species and their rates of extinction, distribution, and protection," *Science* 344, no. 6187 (2014): 1246752.

95 As quoted by Arthur C. Clarke in Andrew Chaikin, "Meeting of the Minds: Buzz Aldrin Visits Arthur C. Clarke," *Space Illustrated*, February 27, 2001.

96 Oliver Milman, "Al Gore: Climate Change Threat Leaves 'No Time to Despair' over Trump Victory," *Guardian* (US), website of the *Guardian* (UK), last modified December 5, 2016, www.theguardian.com/us-news/2016/dec/05/al-gore-climate-change-threat-leaves-no-time-to-despair-over-trump-victory.

## ABOUT THE AUTHOR

Kenneth Lacovara has unearthed some of the largest dinosaurs ever to walk our planet, including the supermassive *Dreadnoughtus*, which at 65 tons weighs more than seven *T. rex*. Through his work, blending exploration with the latest techniques from medicine and engineering, Lacovara portrays dinosaurs as vigorous, competent creatures—the adaptable champions of an age.

Lacovara is sought around the world for his ability to bring the wonders of science and the thrill of discovery to a wide range of audiences. "He's got a way of illuminating the bigger picture, of somehow turning 'why' into wonder. It's not just that he's speaking about what happened way back when, but what it might mean for us today," says Chris Anderson, Curator of TED.

He has appeared in many television documentaries and his discoveries have landed him three times in *Discover* magazine's 100 Top Science Stories of the year and in *Time*'s Top Stories of 2014. Lacovara was named by *Men's Journal* as one of "The Next Generation of Explorers" and he is an elected fellow of the prestigious Explorers Club in New York.

Kenneth Lacovara is the founder and director of the Jean and Ric Edelman Fossil Park of Rowan University in New Jersey.

## WATCH KENNETH LACOVARA'S TED TALK

Kenneth Lacovara's TED Talk, available for free at TED.com, is the companion to *Why Dinosaurs Matter*.

PHOTO: BRET HARTMAN/TED

## RELATED TALKS

**Carrie Nugent**
***Adventures of An Asteroid Hunter***
TED Fellow Carrie Nugent is an asteroid hunter — part of a group of scientists working to discover and catalog our oldest and most numerous cosmic neighbors. Why keep an eye out for asteroids? In this short, fact-filled talk, Nugent explains how their awesome impacts have shaped our planet, and how finding them at the right time could mean nothing less than saving life on Earth.

**Peter Ward**
***A Theory of Earth's Mass Extinctions***
Asteroid strikes get all the coverage, but "Medea Hypothesis" author Peter Ward argues that most of Earth's mass extinctions were caused by lowly bacteria. The culprit, a poison called hydrogen sulfide, may have an interesting application in medicine.

**Nizar Ibrahim**
***How We Unearthed the Spinosaurus***
A 50-foot-long carnivore who hunted its prey in rivers 97 million years ago, the Spinosaurus is a "dragon from deep time." Paleontologist Nizar Ibrahim and his crew found new fossils, hidden in cliffs of the Moroccan Sahara desert, that are helping us learn more about thc first swimming dinosaur—who might also be the largest carnivorous dinosaur of all.

**Al Gore**
***The Case for Optimism on Climate Change***
In this spirited talk, Al Gore asks three powerful questions about the man-made forces threatening to destroy our planet—and the solutions we're designing to combat them.

MORE FROM TED BOOKS

**Asteroid Hunters**
by Carrie Nugent

Everyone's got questions about asteroids. What are they and where do they come from? And most urgently: Are they going to hit earth? Asteroid hunter Carrie Nugent reveals everything we know about asteroids, and how new technology may help us prevent a natural disaster.

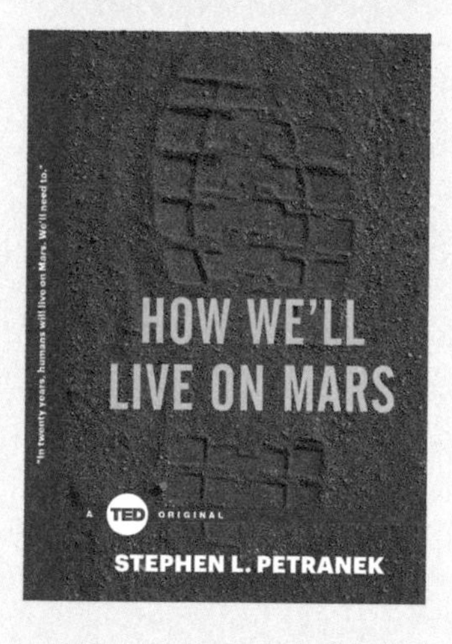


**How We'll Live on Mars**
by Stephen Petranek

It sounds like science fiction, but award-winning journalist Stephen Petranek considers it fact: Within 20 years, humans will live on Mars. We'll need to. In this sweeping, provocative book that mixes business, science, and human reporting, Petranek makes the case that living on Mars is an essential backup plan for humanity, and explains in fascinating detail just how it will happen.

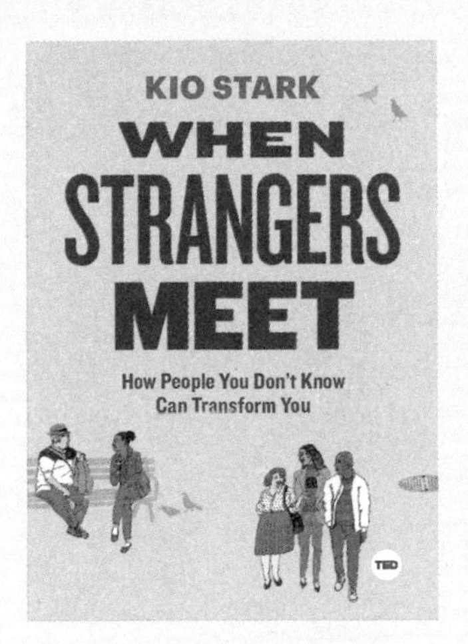


**When Strangers Meet:**
***How People You Don't Know Can Transform You***
by Kio Stark

*When Strangers Meet* reveals the transformative possibility of talking to people you don't know—how these beautiful interruptions in daily life can change you and the world we share. Kio Stark argues for the surprising pleasures of talking to strangers.

David Rothkopf
THE
GREAT
QUESTIONS
*of*
*TOMORROW*
TED

**The Great Questions of Tomorrow**
by David Rothkopf

We are on the cusp of a sweeping revolution—one that will change every facet of our lives. The changes ahead will challenge and alter fundamental concepts such as national identity, human rights, money, and markets. In this pivotal, complicated moment, what are the great questions we need to ask to navigate our way forward?

## ABOUT TED BOOKS

TED Books are small books about big ideas. They're short enough to read in a single sitting, but long enough to delve deep into a topic. The wide-ranging series covers everything from architecture to business, space travel to love, and is perfect for anyone with a curious mind and an expansive love of learning.

Each TED Book is paired with a related TED Talk, available online at TED.com. The books pick up where the talks leave off. An 18-minute speech can plant a seed or spark the imagination, but many talks create a need to go deeper, to learn more, to tell a longer story. TED Books fill this need.

## ABOUT TED

TED is a nonprofit devoted to spreading ideas, usually in the form of short, powerful talks (eighteen minutes or less) but also through books, animation, radio programs, and events. TED began in 1984 as a conference where Technology, Entertainment, and Design converged, and today covers almost every topic—from science to business to global issues—in more than 100 languages. Meanwhile, independently run TEDx events help share ideas in communities around the world.

TED is a global community, welcoming people from every discipline and culture who seek a deeper understanding of the world. We believe passionately in the power of ideas to change attitudes, lives, and, ultimately, our future. On TED.com, we're building a clearinghouse of free knowledge from the world's most inspired thinkers—and a community of curious souls to engage with ideas and each other, both online and at TED and TEDx events around the world, all year long.

In fact, everything we do—from the TED Radio Hour to the projects sparked by the TED Prize, from the global TEDx community to the TED-Ed lesson series —is driven by this goal: How can we best spread great ideas?

TED is owned by a nonprofit, nonpartisan foundation.